Praise for *Two Percent Solutions for the Planet*

"An excellent read and source of ideas for farmers anywhere. With agriculture today producing twenty times as much dead, eroding soil as food required per human each year, it is time to return to farmer creativity. Courtney White's book shows us how creative and observant farmers and ranchers are finding solutions to many of the most challenging problems we face."

—ALLAN SAVORY, president, Savory Institute; chairman,
Africa Centre for Holistic Management

"Courtney White has written one of the most important books for our time about how to reverse climate change and nourish the world with abundant, healthy food. Most importantly, he describes proven, 'shovel-ready' solutions that progressive ranchers and farmers are doing every day. There is no need to spend billions of dollars on new, unproven technologies such as carbon capture and geoengineering. For a fraction of the cost, the world easily could scale up the regenerative practices profiled in *Two Percent Solutions for the Planet*. Too good to be true? Read this book and make up your own mind."

— ANDRÉ LEU, president, IFOAM-Organics International;
author of *The Myths of Safe Pesticides*

"Taking soil seriously offers real leverage in the climate change battle. For those who eat (or raise) meat, *Two Percent Solutions for the Planet* offers fascinating new insights about animal agriculture. For those who eat lower on the food chain, Courtney White details dozens of other ways to help restore the degraded landscapes that, sadly, dominate much of our planet."

—BILL McKIBBEN, author of *Deep Economy*

"Courtney White's *Two Percent Solutions for the Planet* features good sense paired with 50 solutions to our planet's ills, especially those related to carbon. While each solution could be a book in itself, these short profiles may well be the way many people first discover that such solutions are in the offing. It's a brilliant way to inform readers so that they'll prick up their ears when they encounter possible actions based on solutions such as these. This accessible, hope-filled, and beautifully crafted book should be in every school—if not simply everywhere."

—DEBORAH MADISON, author of *Vegetable Literacy*

"This book is Courtney White's most important work. It is the best practical guide to how we can begin to address the significant, unavoidable challenges awaiting us in our not-too-distant future. *Two Percent Solutions for the Planet* inspires us to address these challenges creatively, especially with respect to our food and agriculture future, and to do it in cooperation with nature in ways that also heal the planet. The solutions Courtney describes are not just ideas but are demonstrated strategies already being implemented by creative farmers, ranchers, ecologists, and designers. This book is a must-read for anyone interested in practical ways to restore the planet's health while experiencing a flourishing life."

—FREDERICK KIRSCHENMANN, author of
Cultivating an Ecological Conscience

"The two percents really add up in this glorious presentation of how to make ranching, agriculture, and the great outdoors a major force for addressing our gravest threat: climate change. Courtney White truly knows these solutions, having helped pioneer them as leader of a coalition of ranchers and conservationists."

—GUS SPETH, author of *Angels by the River*; former dean,
Yale School of Forestry and Environmental Studies

"The problems that humanity faces today are the sum total of billions of small missteps. Courtney White focuses on the solutions that will arise from billions of small right steps, and the most important step is the next one that each of us takes. We may never convince seven billion souls to plant one tree at a time, over and over for the rest of their lives, but each one of us can make that commitment if we choose. *Two Percent Solutions for the Planet* shows us a broad pattern for healing the earth that is elegant in its simplicity. We can reforest our world, restore grasslands, sequester carbon, build soil, purify water, provide wildlife habitat, feed humanity, and improve health and nutrition while creating the ecological abundance of the future. Bravo, Courtney!"

—MARK SHEPARD, author of *Restoration Agriculture*

"Courtney White chronicles a new and critically important sphere of knowledge: a world of soil, sun, sky, and animals where good people regenerate the earth in ancient and novel ways. Reading about the environment rarely brings one as many smiles and as much joy as *Two Percent Solutions for the Planet*."

—PAUL HAWKEN, author of *Blessed Unrest*;
coauthor of *Natural Capitalism*

Two Percent Solutions
for the Planet

ALSO BY COURTNEY WHITE

Two Percent Solutions for the Planet

50 Low-Cost, Low-Tech, Nature-Based Practices for Combatting Hunger, Drought, and Climate Change

Courtney White

A **QUIVIRA** COALITION project

Chelsea Green Publishing
White River Junction, Vermont

Photograph of beaver on pages 175, 179, 183, 187, 191, 195, 199, 203, 207, and 211 courtesy of Kkimages/Shutterstock.
Photograph of windmill on pages 91, 95, 99, 103, 107, 111, 115, 119, 123, and 127 courtesy of Rafal Olkis/iStockphoto.

Some of the profiles in this book are adapted from material published in the following books and magazines: *Revolution on the Range* (Island Press, 2008); *Grass, Soil, Hope* (Chelsea Green, 2014); *The Age of Consequences* (Counterpoint Press, 2015); *Farming* magazine; *Acres* magazine; and *Resilience* (journal of the Quivira Coalition). Grateful acknowledgment is given to these publications.

Project Manager: Alexander Bullett
Acquisitions Editor: Benjamin Watson
Project Editor: Fern Marshall Bradley
Copy Editor: Susan Davidson
Proofreader: Eric Raetz
Indexer: Linda Hallinger
Designer: Melissa Jacobson

Printed in the United States of America.
First printing September, 2015.
10 9 8 7 6 5 4 3 2 1   15 16 17 18

Our Commitment to Green Publishing
Chelsea Green sees publishing as a tool for cultural change and ecological stewardship. We strive to align our book manufacturing practices with our editorial mission and to reduce the impact of our business enterprise in the environment. We print our books and catalogs on chlorine-free recycled paper, using vegetable-based inks whenever possible. This book may cost slightly more because it was printed on paper that contains recycled fiber, and we hope you'll agree that it's worth it. Chelsea Green is a member of the Green Press Initiative (www.greenpressinitiative .org), a nonprofit coalition of publishers, manufacturers, and authors working to protect the world's endangered forests and conserve natural resources. *Two Percent Solutions for the Planet* was printed on paper supplied by Quad-Graphics that contains at least 10% postconsumer recycled fiber.

Library of Congress Cataloging-in-Publication Data
White, Courtney, 1960-
 Two percent solutions for the planet : 50 low-cost, low-tech, nature-based practices for combatting hunger, drought, and climate change / Courtney White.
    pages cm
 Includes bibliographical references and index.
 ISBN 978-1-60358-617-7 (pbk.)—ISBN 978-1-60358-618-4 (ebook)
1. Environmental protection—Citizen participation. 2. Sustainable living. I. Title. II. Title: 2% solutions for the planet.

 TD171.7.W453 2015
 363.7'06--dc23

                    2015018532

Chelsea Green Publishing
85 North Main Street, Suite 120
White River Junction, VT 05001
(802) 295-6300
www.chelseagreen.com

"The only progress that counts is that on the actual landscape of the back forty."

—ALDO LEOPOLD

Contents

PART FOUR

Restoration

PART FIVE

Wildness

Solutions Abound

We live in an era of seemingly intractable challenges: increasing concentrations of carbon dioxide (CO_2) in the atmosphere, rising food demands from a human population that is projected to expand from seven to nine billion people by 2050, and dwindling supplies of fresh water, to name just three. What to do? So far, our response to these big problems has been to consider "big" solutions, including complex technologies, arm-twisting treaties, untested geoengineering strategies, and new layers of regulation, all of which have the net effect of increasing complexity (and anxiety) in our lives. And most of these big solutions come with big costs, both financial and social, especially for those least able to bear them.

Which raised a question in my mind a few years ago: Why not consider low-cost, low-tech, nature-based solutions instead?

I knew this was possible based on my experience with the Quivira Coalition, a New Mexico–based nonprofit that I cofounded in 1997 with a cattle rancher and a fellow conservationist. Our original goal was to find common ground between ranchers, conservationists, public land managers, scientists, and others around progressive livestock grazing practices that were good for both the land and its inhabitants. Over time our work increasingly focused on building economic and ecological resilience in the West, with a special emphasis on ecological restoration, local food production, and bridging urban-rural divides (described my book *Revolution on the Range*).

Through Quivira, I had met many innovative people who had been hard at work for decades field-testing and implementing a wide variety of regenerative land management practices, proving them to be practical, profitable, and effective. These practices, such as planned grazing by livestock and the ecological restoration of creeks, are principally low-tech, involving photosynthesis, water, plants, animals, and thoughtful stewardship rather than big-ticket technological interventions. I knew they improved land health, produced food, and repaired broken water cycles. What I didn't know was how they might address the rising challenge of greenhouse gas buildup in our atmosphere.

This changed in 2009 when a Worldwatch Institute report, "Mitigating Climate Change through Food and Land Use," landed on my desk.

Its authors argued that the potential for removal of CO_2 from the atmosphere through plant photosynthesis and related land-based carbon sequestration activities was both large and largely overlooked. Strategies they listed included enriching soil carbon, no-till farming with perennials, employing climate-friendly livestock practices, conserving natural habitat, and restoring degraded watersheds and rangelands.

That sounded like the work of the Quivira Coalition!

Exploring further, I discovered that many other regenerative practices also sequester CO_2 in soils and plants as well as address food and water problems. The link, I learned, was carbon. It's the soil beneath our feet, the plants that grow, the land we walk, the wildlife we watch, the livestock we raise, the food we eat, the energy we use, and the air we breathe. Carbon is the essential element of life. A highly efficient carbon cycle captures, stores, releases, and recaptures biochemical energy, making everything go and grow from the soil up. A healthy carbon cycle, I realized, had a wide range of positive benefits for every living thing on the planet.

However, I also discovered that carbon sequestration in soils and the climate change mitigation potential of these regenerative and resilient practices was nearly unknown to the general public, much less to decision makers and others in leadership positions. Even within progressive ranching, farming, and conservation communities, the multiple economic and ecological gains that come from increasing carbon in soils were largely overlooked. The story of carbon needed to be told, I saw, leading me to write *Grass, Soil, Hope: A Journey through Carbon Country*, which makes the case that if we can draw increasing amounts of carbon out of the atmosphere and store it safely in the soil we can go a long way toward solving many of the challenges that now confront us.

There wasn't enough space in *Grass, Soil, Hope* for many of the hopeful stories of regenerative practices that I had discovered along the way. What to do with all these wonderful solutions? After giving it some thought and consulting with my colleagues at the Quivira Coalition, I decided to begin writing them up as short case studies. There was a need, I surmised, for succinct profiles of nature-based approaches to global problems. To that end, I included some condensed versions of practices I had described in *Grass, Soil, Hope* and other publications. I called the entire series *2% Solutions for Hunger, Thirst and CO_2* and we bundled 14 profiles into a special edition of Quivira's journal, *Resilience*, in the fall of 2013. The response was very positive, so I decided to keep writing—resulting in this book: *Two Percent Solutions for the Planet*.

The 2 percent in the title refers to:

- the small amount of additional carbon in the soil needed to reap a wide variety of ecological and economic benefits;
- the portion of the nation's population who are farmers, ranchers, and others who can get this work done; and
- the low financial cost of these solutions—only 2 percent of the nation's gross domestic product.

It is an illustrative number—not a scientific one—meant to stimulate our imaginations. *Look what can happen for only 2 percent!* Big solutions, in other words, can actually be accomplished at a small cost.

Each of the fifty practices profiled in this book either builds soil carbon (and thus mitigates climate change), intensifies food production sustainably, improves water quality and quantity, or involves a critical support activity—such as animal herding or an advance in appropriate technology—that enhances regenerative practices. Many do more than one! Some solutions are more expensive, complicated, or specialized than others, but all share a common attribute: they are regenerative over the long haul, meaning they *replete* rather than *deplete* people, animals, plants, soil, and other natural resources. Each solution is simultaneously unique and interdependent. Each can be implemented on its own, depending on local conditions and circumstances, but each is also part of a synergistic *whole*—a vision of renewability, vitality, and careful stewardship. You can use them like tools lifted from a toolbox, but without a larger blueprint in mind you won't build anything durable.

In the last decade or two, a movement to rediscover and implement "old" practices of bygone days has arisen rapidly, abetted by remarkable innovations in technology; breakthroughs in scientific knowledge; and tons of old-fashioned, on-the-ground problem-solving. Some of the reasons for the rapid development of this "new" agriculture are practical; some are economic; some are philosophical; and some are driven by a sense of urgency about the world—but all of them are motivated by a concern for the future. Questions abound: How can we conserve finite and dwindling natural resources for future generations? How can we adjust and adapt our lives to tomorrow's changing climate? How can we create a robust economic and ecological bequest for all our children?

To find answers, many people looked to the past for wisdom, and what they discovered is this: nature's model works best. After all, nature has used evolution and the laws of physics to beta-test what works for merely millions of years—billions in the case of photosynthesis. That's why a new generation of agrarians is returning to the roots of agriculture and conservation for a different approach, with large helpings of science and social advancement added in. I like the way the Rodale Institute described it recently in a white paper: *farming like the Earth matters*. Like water and soil and land matter. Like clean air matters. Like human health, animal health, and ecosystem health matter.

It all matters, and regenerative solutions are the way we'll get there.

The goal of this book is to present informative snapshots of regenerative practices in a format that can be widely read and shared. It is not a comprehensive accounting by any means. I picked 50 topics that I consider to be a diverse representation of the regenerative world. There are other solutions already at work, and new ones are being developed even as you read this. I encourage you to seek them out. In the meantime, I hope this book will help you connect the dots between these diverse, pragmatic, and hopeful practices.

It is also my hope that readers will be energized by a story or two in the collection to explore a particular topic further—to dig deeper and learn more. Consider each solution as the top two inches of water in a well that extends down hundreds or thousands of feet—a well of knowledge and experience that took decades to create. Check out these wells in the field yourself: visit a farm or ranch or research project and learn directly from the practitioners themselves.

This collection has one more goal: spread the good news—solutions abound!

TO LEARN MORE

"Mitigating Climate Change through Food and Land Use"
by Sara J. Scherr and Sajal Sthapit. WorldWatch Report no. 179.
WorldWatch Institute, Washington DC, 2009.

Grass, Soil, Hope: A Journey through Carbon Country
by Courtney White. Chelsea Green Publishing,
White River Junction, VT, 2014.

PART ONE

Ranching

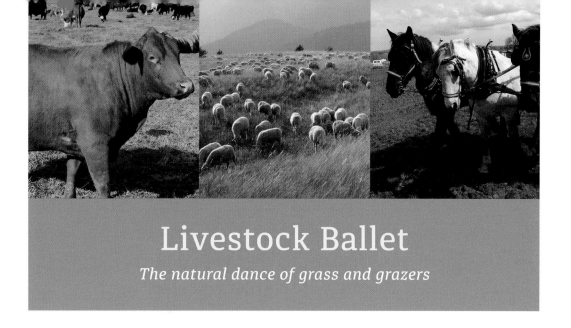

Livestock Ballet

The natural dance of grass and grazers

"Nature never farms without animals" is an old saying in organic agriculture.

Over the past 66 million years, North American grasslands evolved under grazing pressure from herds of ruminants (bison, deer, antelope) and other herbivores. The animals would graze areas for short periods of time, move on to fresh grass, come back later if the conditions were right, and then go away again, in an ancient and sophisticated dance of sunlight, soil, and rain. For its part, grass reciprocates the dance steps. When old, dormant, or dead grass is removed by grazing animals (or burned off by a natural fire) new grass emerges and grows vigorously, as long as there is sufficient snow or rain. Grass and grazers need each other, in other words. The rhythm of the performance changes from year to year and place to place, depending on the needs of the dancers, but when there was enough moisture the result was always the same: a photosynthetic standing ovation.

The holistically minded ranchers and farmers that I've had the honor to meet, most of whom credit pioneering biologist Allan Savory for their inspiration, know this dance by heart and have become choreographers on their own properties, mimicking nature as closely as possible. The dancers (wild and domesticated) might be different, but the goal and the basic dance steps are the same: healthy land and the ecological processes that sustain it. Like any good performance, the key is timing—in this case how long the grazing lasts in any one spot. Too much and the land suffers, too little and the grass struggles to reach its potential. Call it the Goldilocks principle: the animals shouldn't stay too long or come back too early—everything has to be just right.

For some ranchers and farmers, the dance is a quiet pas de deux, but for others it's like having two or three different marching bands constantly rotating on and off the stage, requiring more complicated choreography. One such dance master is Joel Salatin, a well-known maverick farmer and evangelist for agroecological practices and profits. On his family's Polyface Farm, located in western Virginia, cattle, chickens, and pigs—even rabbits and turkeys—are carefully rotated across the farm's 550 acres in what Salatin calls a "livestock ballet."

It's an annual performance that earns rave reviews. In his best-selling book *Omnivore's Dilemma*, Michael Pollan admiringly describes Salatin's ability "to choreograph the symbiosis of several different animals, each of which has been allowed to behave and eat as they evolved" and thus nearly eliminate the need for machinery, fertilizers, chemicals, or mechanical waste disposal—all of which has important positive implications for the carbon content of the farm.

The ballet starts in the barn in winter, where Polyface's cattle are fed hay harvested during the previous growing season. Unlike other farms, however, the manure isn't shoveled out. Instead, it's covered with wood chips and straw every few days and "salted" with corn. As the compost pile grows, the heat generated keeps the cattle warm while fermenting the corn. When the cows turn out in the spring, several dozen pigs are brought into the barn to do their thing: dig up the compost with their noses. Pigs love fermented corn, and as a result the compost pile is thoroughly aerated by their rooting. This process transforms compacted, anaerobic (oxygen-less) dirt into fluffy, aerobic (oxygen-rich) soil, full of biological life.

These are Salatin's famous "pigerators" at work, employing what he calls the "pigness of the pig" to get the job done. When the pigs are finished, Salatin spreads the carbon-rich compost on Polyface's pastures, where it feeds the microbes that will feed the grass that feeds the cows. Next in the ballet, Salatin grazes the cattle as a herd in small paddocks ringed by electric fencing, often for only a single day before moving them to fresh grass. Employing the "cowness" of the cow, Salatin calls this procedure "mobbing, mowing, and moving."

"Herbivores in nature exhibit three characteristics," he said in an interview with *Acres* magazine, "mobbing for predator protection, movement daily onto fresh forage and away

Innovative farmer Joel Salatin employs his famous pigerators to help turn cow manure into high-quality compost. *Photo courtesy of Polyface Farms*

from yesterday's droppings, and a diet consisting of forage only—no dead animals and no grain. . . . This natural model heals the land, thickens the forage, reduces weeds, stimulates earthworms, reduces pathogens, and increases nutritional qualities in the meat."

The next performers in the Polyface ballet are chickens. Five hundred hens are brought into a field three days after the cows leave. They arrive in an "eggmobile"—a large covered coop on four wheels—and proceed immediately to express their "chickenness" by scratching apart the manure patties left behind by the cattle. This sanitizes and recycles the waste very efficiently. The hens also consume lots of bugs, including crickets and grasshoppers, in fields prepped by the cows who have sheared the grass short for

A typical eggmobile moves the flock daily so the hens can range on fresh pasture (a farm in New Hampshire in this case). *Photo by Courtney White*

them. All the while, the chickens are producing lots of healthy eggs. It's a similar process for Polyface's broiler chickens, rabbits, and turkeys, all of which are shuffled around the farm in small portable pens placed directly on the ground and moved every day to fresh forage.

As Salatin likes to point out, this is how potential liabilities (animal waste disposal) are turned into profitable assets—organic, grassfed food in this case.

In another part of the ballet, the pigs are released into Polyface's woods for a month or more, during which they root for food in the soil, creating a healthy disturbance in the forest floor. Trees are part of the dance too. When left uncut, they soak up atmospheric carbon dioxide (CO_2), thanks to photosynthesis; when cut, they provide firewood for the farm and wood chips for the compost. The woods also shelter songbirds, which eat bugs and provide prey for predators, which then leave Salatin's chickens alone.

It's all part of a holistic vision for the farm.

"The carbon from the woodlots feeds the fields," Pollan wrote about Polyface, "finding its way into the grass and, from there, into the beef. Which it turns out is not only grass fed but tree fed as well . . . a hundred acres of productive grassland patchworked into four hundred and fifty acres of unproductive forest. It was all of a biological piece, the trees and the grasses and the animals, the wild and the domestic, all part of a single ecological system."

It's a system that is very good at creating topsoil, which in turn can soak up lots of CO_2. During photosynthesis carbon (C) is separated from oxygen (O_2), and a lot of this carbon makes its way underground via plant roots, where it can safely be stored for long periods of time. The key is promoting and maintaining the dance of life in the soil, which requires beautiful music created by an orchestra of animals, humans included. When everything harmonizes, the effect can be amazing.

"This is all extremely symbiotic," said Salatin in the *Acres* interview, "and creates a totally different relationship than when you're simply trying to grow the fatter, bigger, cheaper animal."

In other words, just like a dance, farming and ranching done right are all about diverse, strong, and reciprocal relationships. We need each other—grass, grazers, eaters, producers, the domestic, the untamed, the dance steps, and the music.

"Relations are what matter most," Pollan summed up, "and the health of the cultivated turns on the health of the wild."

TO LEARN MORE

Joel Salatin's publications and videos are available on the Polyface Farm website: www.polyfacefarms.com

Holistic Management: A New Framework for Decision Making by Allan Savory with Jody Butterfield. Island Press, Washington, DC, 1999.

Healing Ground

Restoring the carbon cycle with cattle

C attle have an important role to play in fighting climate change.

It starts with a key concept: the ecological processes that sustain wildlife habitat, biological diversity, and functioning watersheds are the same processes that make land productive for livestock. It's called *land health*—the degree to which the integrity of an ecosystem is sustained over time. Before land can sustainably support a human value, such as livestock grazing, hunting, recreation, or wildlife protection, it must be functioning properly at a basic ecological level, which includes healthy water, mineral, and carbon cycles flowing round and round from the atmosphere to soil to plants to animals and back again. The trouble is a great deal of land around the world is in poor health, mostly a result of lousy management, and is in need of healing.

Which is where cattle come in, as I'll explain with a real-world example.

In 2004, Tom and Mimi Sidwell bought the 7,000-acre JX Ranch, south of Tucumcari, New Mexico, and set about doing what they know best: earning a profit by restoring the land to health and stewarding it sustainably.

As with many ranches in the arid Southwest, the JX had been hard used over the decades. Poor land management had caused the grass cover to diminish in quantity and quality, exposing soil to the erosive effects of wind, rain, and sunlight and significantly diminishing the organic content of the soil, especially its carbon. Eroded gullies had formed across the ranch, small at first but growing larger with each thundershower, cutting down through the soft soil, biting deeper into the land, eating away at its vitality. Water tables fell correspondingly, starving plants and animals alike of precious nutrients and forage.

Profits fell too for the ranch's previous owners. Many had followed a typical business plan: stretch the land's ecological capacity to the

Rancher Tom Sidwell standing on restored grasslands on the JX Ranch, a result of planned grazing practices and brush clearing. *Photo by Courtney White*

breaking point, add more cattle when the economic times turned tough, and pray for rain when dry times arrived, as they always had. The result was always the same—a downward spiral ecologically and economically. In the end, nutrient, mineral, and energy cycles unraveled across the ranch, causing the land to disassemble and eventually fall apart.

Enter the Sidwells. With thirty years of experience in holistic planned grazing, which controls the timing, intensity, and frequency of livestock grazing across a parcel of land, they saw the deteriorated condition of the JX not as a liability but as an opportunity. Tom Sidwell began by dividing the entire ranch into 16 pastures, up from the original 5, using solar-powered electric fencing. After installing a water system, he picked cattle from breeds that do well in dry country, grouped them into one herd, and set about carefully rotating them through all 16 pastures, never grazing an individual pasture for more than 10 days in order to give the grass plenty of time to recover before being grazed again. This mimics the graze-and-go behavior of wild herbivores, such as bison, who would heavily impact a stretch of land and then move on to fresh grass. Many ranchers, in contrast, allow their cattle to graze a stretch of land continuously.

Next Sidwell began clearing out the juniper and mesquite trees on the ranch with a bulldozer, which allowed native grasses to come back. As grass returned, Sidwell lengthened the period of rest between pulses of grazing in each pasture from 60 to 105 days across the whole ranch. More rest meant more time for grass to grow which, in turn, meant he could eventually graze more cattle. Round and round, spiraling upward ecologically, instead of downward. In fact, over the past 10 years the JX has seen an increase in diversity of grass species, including cool-season grasses, and a decrease in the amount of bare soil across the ranch, which has allowed Sidwell to increase the livestock capacity of the JX by 25 percent, significantly impacting the ranch's bottom line.

Sidwell considers maintaining soil health to be the key to the ranch's success. To do that he plans his grazing sequences so that standing vegetation and litter remain on the soil surface even after the cattle have grazed. The surface covering decreases the impact of raindrops on bare soil, slows runoff to allow water infiltration, provides cover for wildlife, and feeds the microorganisms in the soil. He also plans ahead for drought by adjusting his livestock numbers *before* the drought takes off, instead of during or after the drought has set in, as is traditional.

"I plan for the drought," Sidwell said with a wry smile, "and so far, everything is going according to plan."

There is an important collateral benefit to all this planning: the Sidwells' cattle are healing the carbon cycle. Here's why: photosynthesis transforms sunlight and CO_2 into biochemical energy (glucose), which is resynthesized into a variety of carbon compounds, some of which make their way into the soil via roots and eventually form humus—a chemically stable type of organic matter that most people associate with rich gardens. Building humus pulls CO_2 out of the atmosphere and sequesters the carbon in soil for long periods of time. However, decomposing organic matter and active microorganisms in the soil "breathe out" CO_2 (completing the carbon cycle), which means it's critical to have more CO_2 "breathed in" via green plants than gets released.

If land is bare, degraded, or unstable due to erosion and if it can be restored to a healthy condition, with properly functioning carbon, water, mineral, and nutrient cycles, and covered with green plants with

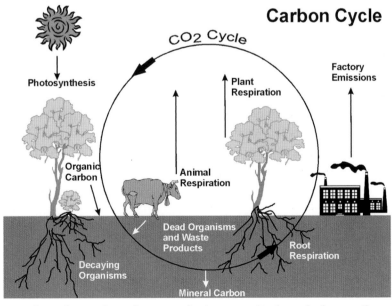

How the carbon cycle works: atmospheric CO_2 cycles through plants, soil, and animals, and back into the air again. Extra CO_2 is created by the burning of fossil fuels. *Image by Tamara Gadzia for the Quivira Coalition*

deep roots, then the quantity of CO_2 that can be sequestered is potentially high. Conversely, when healthy, stable land becomes degraded or loses green plants, the carbon cycle can become disrupted and will release stored CO_2 back into the atmosphere. It's an ancient equation: more plants mean more green leaves, which mean more roots, which mean more carbon exuded by those roots, which means more CO_2 can be sequestered in the soil, where it will stay.

Which is exactly what the Sidwells are doing.

There's another benefit to carbon-rich soil: due to its spongelike quality it improves water infiltration and storage. Recent research indicates that one part carbon-rich soil can retain as much as four parts water. This has important positive consequences for the recharge of aquifers and base flows to rivers and streams, which are the lifeblood of towns and cities.

It's also important to people who make their living off the land, as Tom and Mimi Sidwell can tell you. Recently they were pleased to discover that a spring near their house had come back to life. For years it had flowed at the miserly rate of one-quarter gallon per minute, but after the Sidwells cleared out the juniper trees above the spring and managed the cattle for increased grass cover, the well began to pump water at six times that rate, 24 hours a day.

Improving the water cycle has a direct benefit for another reality in arid lands: drought. In 2011, less than 3 inches of rain fell on the JX over a period of 12 months. In response, Sidwell asked himself, "What would a bison herd do?" They would avoid a droughty area, he decided, so he sold almost the entire cattle herd in order to give his grass a rest. It was a gamble, but it paid off in 2012 when it began raining again, although the total amount was 10 inches below normal.

"It was enough to make a little grass," Sidwell said, "but I think the roots are strong and healthy and recovery will be quick. Grazing and drought planning are a godsend."

So is soil carbon!

TO LEARN MORE

Visit the JX Ranch website: www.leannaturalbeef.com

Cows Save the Planet: And Other Improbable Ways of Restoring Soil to Heal the Earth by Judith D. Schwartz. Chelsea Green Publishing, White River Junction, VT, 2013.

Why Grassfed

Best for you, best for the planet

I
f there's one overarching lesson I've learned over the years, it's this: nature knows best.

Take meat, for instance. The most frequent answer I hear at conferences when a distraught member of the audience asks a presenter, "What's the one thing I can do for the planet?" is the instruction to eat less red meat. It's an understandable response, but what should have been said is, "Eat less *feedlot* meat." Actually, the correct answer is to eat *grassfed* meat (if one is inclined to eat meat at all). By *grassfed* I mean meat from animals that have spent their entire lives on grass. It's how nature intended ruminants, such as cattle, and other herbivores to live—as grassfed.

Unfortunately, humans think they know best, which is how we ended up with a food system that provides more than 90 percent of our beef from crowded, stinky, grassless industrial feedlots where livestock subsist on corn and associated by-products that they were not designed by nature to eat. The sins of the feedlot system have been well documented, beginning very publicly in 2002 when the *New York Times Magazine* published Michael Pollan's exposé "Power Steer." By following a steer ("No. 534") from cattle ranch to feedlot to slaughter, Pollan discovered a disturbing list of industrial troubles, including the sickness and abuse animals must endure; the air, land, and water pollution caused by these operations; the deleterious use of hormones and antibiotics needed to keep the animals alive; the low pay and stressful conditions for feedlot workers; the reduction in nutritional content of the meat; and the drenching of the entire system in fossil fuels.

"The only big advantage of feedlot beef," wrote Pollan, "is that it's remarkably cheap." While that makes economic sense—sort of—it makes no ecological sense. Pollan voted for grassfed beef. "Eating a

Cattle that spend their entire lives on pasture are called grassfed, which is how nature intended them to live. *Photo courtesy of Morris Grassfed Beef*

steak at the end of a short, primordial food chain," he concluded, "comprising nothing more than ruminants and grass and light is something I'm happy to do and defend."

Around the same time that Pollan's article appeared, author and researcher Jo Robinson began to write about the emerging scientific evidence supporting the health benefits of grassfed over corn-fed feedlot meat for humans, which she summarizes as:

- More omega-3 fatty acids ("good" fats) and fewer omega-6 ("bad" fats);
- Lower in the saturated fats linked with heart disease;
- Much higher in conjugated linoleic acid (CLA), a cancer fighter;
- Much more vitamin A;
- Much more vitamin E;
- Higher in beta-carotene;
- Higher in the B vitamins thiamin and riboflavin;
- Higher in calcium, magnesium, and potassium;
- No traces of added hormones, antibiotics, or other drugs.

As Robinson likes to say, "If it's in their feed, it's in our food"—which means it's in us. This is a very important reason why grassfed products are best, including grassfed milk, cheese, and other dairy products. As for eating less red meat, Robinson said in an interview posted on her website, "I'm not one of those who think that eating less meat is good.

I think eating less of the wrong kind of meat is very good and very important. I think we can have up to 40 percent of our calories from meat, and that's fine as long as it's healthy meat."

Best of all, if grassfed meat and dairy are produced as part of a holistically managed farm or ranch that employs nature-based grazing practices, then it truly becomes a win-win for people and the planet.

A good example is Morris Grassfed Beef. In 1991, Joe Morris became one of the first ranchers in California to offer grassfed beef to customers, predating the recent boom in grassfed production by a dozen years. Born and raised in San Francisco, Morris was inspired to give livestock a go by his grandfather, who owned and ran a ranch near San Juan Bautista, south of San Jose. Inspired by the writings of Wendell Berry, Morris decided to reject the industrial model of production in favor of a type of agriculture that worked with nature's principles. When he discovered the holistic grazing practices championed by Allan Savory, which mimic the natural grazing behavior of wild herbivores, everything fell into place.

In marketing their beef, Morris and his wife, Julie, point out that in addition to the meat's health advantages, good grazing practices have multiple ecological benefits: well-managed pastures absorb far more rainwater than most other agricultural or recreational land uses (very useful in dry times); grazing lands provide open space for wildlife and protect against the encroachment of subdivisions; and nature-based management of cattle can grow deep-rooted perennial plants, which improve nutrient and carbon cycling in the soil. It also proves a good life for a cow or steer—living on pasture grass from birth to its last day as nature intended. You don't have to be a meat eater to appreciate both the humane and environmental benefits that good grazing practices create.

In the past few years, another important advantage of grassfed has emerged: it requires less burning of fossil fuels than feedlot beef—a lot less. By some estimates, meat from grassfed animals requires only one calorie of fossil fuel to produce two calories of food. In contrast, feedlot beef requires five to ten calories of fossil fuel for every calorie of food produced. The main differences are the synthetic fertilizers and other inputs used to grow the corn fed to the cattle in feedlots

Joe and Julie Morris raise grassfed cattle and manage their land with regenerative, nature-based grazing practices. *Photo courtesy of Morris Grassfed Beef*

and the amount of transportation involved in moving feedlot beef across the nation to supermarkets. In contrast, many grassfed meat operations sell their products locally (and their cattle don't eat corn).

It's true that cattle belch methane, a potent greenhouse gas, as a product of the digestive process—and a lot of cattle can produce a lot of methane. However, this becomes another reason to choose grassfed over feedlot beef.

In her book *Defending Beef: The Case for Sustainable Meat Production*, Nicolette Hahn Niman challenges the case that cattle are significant contributors to climate change—at least not the grassfed variety. The amount of CO_2 generated by pastured-meat production is negligible compared to the amount produced by feedlot cattle. As for methane, she notes that nutrition experts have demonstrated that minor changes to cattle diets can cut emissions by as much as 50 percent. She also writes that methane is produced by the biological decomposition of vegetation and will happen regardless of whether a cow has consumed it or not. Then there's the huge potential of carbon sequestration in soils—it can't happen in feedlots (because there's no grass) but it can happen on well-managed rangelands and farms. It's another reason why grassfed is best—and why society needs to readjust its generally negative attitude toward the lowly bovine.

"Cattle are not, in fact, a climate change *problem* at all," writes Niman, "instead, cattle are actually among the most practical, cost-effective *solutions* to the warming of the planet."

In the end, everything circles back to the same point: nature knows best. Ruminant animals eating grass all their lives managed according to natural graze-and-go principles is nature at work. But those farmers and ranchers who are practicing these principles need us to exercise our consumer power in their support. Those of us who eat meat need to make the conscious choice to reject feedlot beef and seek out grassfed beef.

After all, it's the best choice for our health, for the welfare of livestock, and for the well-being of the planet.

TO LEARN MORE

For more information on the health benefits of grassfed beef, visit Jo Robinson's website: www.eatwild.com

Defending Beef: The Case for Sustainable Meat Production by Nicolette Hahn Niman. Chelsea Green Publishing, White River Junction, VT, 2014.

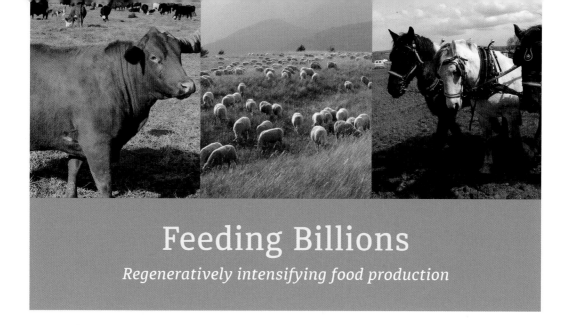

Feeding Billions

Regeneratively intensifying food production

The human population of the planet is on course to grow from seven billion to nine billion by 2050, raising a *huge* question: How are we going to feed an extra two billion people without destroying what's left of the natural world, especially under the stress of climate change?

It's not simply about the planet's poor either. The food well-fed Americans eat comes from a global production system that is already struggling to find enough arable land, adequate supplies of water, and drought-tolerant plants and animals to feed seven billion people. Add two billion more and you have a recipe for a devastating raid on the natural world. Where is all this additional food and water going to come from?

Industry has an answer: more of the same. More chemicals, fertilizers, GMOs, monocropping, and heavy fossil fuel use. In fact, a second global "Green Revolution" is required, they say, even though the consequences of the first one are now haunting us in a variety of ways, including heavy dependence on pesticides and herbicides. Sure, we feed many more people today than we did 60 years ago, but at a high cost to the health of land, animals, and people. More of the same looks like a form of insanity.

Fortunately, a growing body of evidence supports an alternative approach to food intensification: expanding support for smallholder farmers who employ nature-based practices that restore, promote, and sustain the regeneration of life, both above and below the surface of the soil. Rather than "get big or get out"—which has been the mantra of American agriculture for decades—or expand industrial methods of production, or embark on another foray into the murky terrain of GMO technology, why not simply double what we know already works regeneratively?

Take the example of Sam Montoya, who ran 220 head of cattle on only 93 acres of land profitably and sustainably for years. Let me repeat that: *220 head of cattle on only 93 acres of land profitably and sustainably.* When I first heard about Montoya's little farm, located on Sandia Pueblo, a Native American reservation just north of Albuquerque, New Mexico, I was astonished. In the arid Southwest, that many cattle typically need 10,000 acres or more to be managed properly. The difference, of course, is water. Sam's farm was irrigated, but that only makes his story even more intriguing.

Most irrigated land in the West produces hay or alfalfa for animals, not food for people, and those operations that run cattle do so at a stocking rate of about one cow per acre. Montoya's stocking rate was *more than double that*, which meant that he produced twice as much food on irrigated ground than could be accomplished by conventional management. And his method, as I saw during a visit, sometimes resulted in more grass than his cows could eat. That's how he built up his stocking rate over time—by growing more and more grass regeneratively using the tool of grazing animals.

So how did Montoya double his stocking rate while growing more grass?

After retiring from a career with the Bureau of Indian Affairs, Montoya decided that he wanted to return to his agricultural roots. Upon receiving permission from the tribe to rehabilitate a depleted sod farm, located a short distance from the Rio Grande, he laser-leveled the land, planted a variety of grasses, divided the ground into 33 paddocks

As a result of his progressive management, Sam Montoya could run 220 cattle on only 93 acres of irrigated land. *Photo by Courtney White*

(roughly three acres each) with electric fencing, and installed a cattle watering tank in the center that was accessible from all paddocks. When the last dairy in the area shut down, Montoya scored a supply of manure for fertilizer. Then he covered the fields with irrigation water. When the grass grew, he turned the cattle out.

The animals grazed as a single herd in one paddock for only one day. When the 24-hour period was up, Montoya would walk over from his house in the nearby village, lower a gate in the electric fence, watch the cattle walk into the adjacent paddock, secure the fence when the move was complete, and then go home. The entire process took less than half an hour, meeting Montoya's requirement that he "not work too hard" in his retirement, as he explained to me.

The rotation of the herd through the entire sequence of paddocks took a little more than a month. By carefully managing the irrigation water, Montoya ensured that the grass was ready for another harvesting by the time the cattle herd came back to a particular paddock. And he repeated this cycle year round, even through the winter.

"I'm trying to mimic what the bison did," Montoya told me. "They kept moving all the time." The key was the amount of rest he gave the grass before the cattle came back. "You, me, the land—everything needs a break," he said. "But you shouldn't sit on the sofa all week. Too much rest is as bad as too much work. It's all about balance."

Pursuing that balance, Montoya refused to use pesticides, herbicides, or other chemicals. Eventually he stopped adding manure as well. He recycled everything and wasted nothing. Better yet, other than what

Canada geese also enjoy the lush grass produced by Montoya's mob-grazing strategy. *Photo by Courtney White*

was necessary for delivering and hauling away the cattle, Montoya's operation required no fossil-fuel-dependent machinery either, a fact that pleased the economically minded rancher.

"I don't want anything that rusts, rots, or depreciates," said Montoya, grinning. "Plus, I feel good that I'm not polluting the air."

That included not emitting greenhouse gases (other than cow belches). Although it wasn't part of his goal, I'm certain he was also sequestering CO_2 by building new soil on land that had been recently stripped of it as part of the sod farm. It all meshed together because his operation worked on the original solar power: photosynthesis. In fact, Montoya called himself a "grass farmer," which meant he considered grass to be his principle product, not beef. The cattle were his lawn mowers, as he put it.

Perhaps as importantly, Montoya made money. Profits from the sale of cattle—he was a studious observer of business cycles in the livestock industry—allowed him to quickly pay back the loan he took out to get the farm started. In only a few years he operated in the black, thanks to his very low costs.

"It works pretty well," Montoya said the last time I visited. "It's been pretty good to me. And I know it's been good for the land." Although Montoya is retired now from grass farming, what he accomplished on his 93 acres is a great example of a regenerative practice that builds topsoil, intensifies yields, and conserves the natural environment.

With two billion more people to feed very soon, what Montoya did on his little place is a solution we need to heed.

TO LEARN MORE

For more information on mob grazing and regenerative agriculture, read one of the books by rancher Greg Judy or visit his website: www.greenpasturesfarm.net

A valuable resource on grass farming is the monthly newspaper *Stockman Grass Farmer*: www.stockmangrassfarmer.com

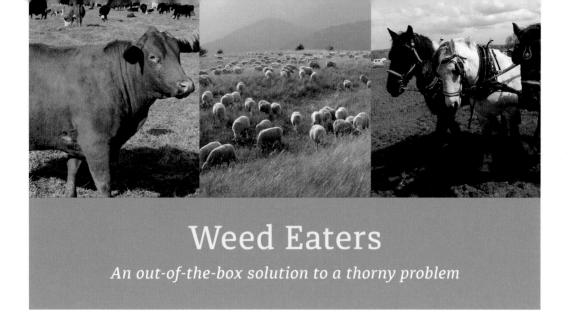

Weed Eaters

An out-of-the-box solution to a thorny problem

W hy use chemistry to solve a problem when you can use biology instead?

In 2004, Kathy Voth came up with a novel strategy to tackle the urgent and rapidly expanding challenge of stopping the spread of invasive weeds across the nation: get cows to eat them. This was way-out thinking because, according to convention at the time, cattle were grazers that strictly ate grass. Goats were browsers that might eat anything, and sheep were something in between. If you wanted livestock to tackle a noxious weed infestation on your farm or ranch, you employed a herd of goats. If you didn't want to use a biological remedy, however, then you could return to the standard solution: costly chemical herbicides—in large quantities. After all, what other practical alternative was there? Not cows.

Yes, cows. Over the past decade, Voth has developed a simple yet effective process for training cows to eat weeds, including almost any kind of cow and almost any type of weed. There's no gimmick involved. Her process is based squarely on recent scientific research into how nature shapes the eating preferences of livestock and on well-established principles of animal behavior. Voth's process takes only 10 hours of training spread over 10 days to teach a group of cattle to eat weeds. It works for a simple reason: the cows never realized weeds tasted so good!

Let's back up for a second. Why worry about invasive weeds?

Over ninety different foreign plants are recognized on the USDA's Federal Noxious Weeds list. Collectively, they infest more than 100 million acres across the country, including 20 percent of our public lands, and they are expanding at a rate of 8 to 12 percent every year (roughly the size of Delaware). Weeds crowd out native plants, damage crops and forage, and contribute to soil erosion. Some can poison wildlife and livestock. Taken together, they are a huge threat, not only to food production but to biodiversity and watershed health as well.

Poor land health can also put human lives at risk. In 1994, Voth was a public information officer in Colorado when a forest fire killed fourteen firefighters. A century of fire-suppression policies by federal and state agencies combined with unusually wet years in the 1970s and '80s set the stage for catastrophic fires across the West. The Colorado tragedy set Voth to thinking about the danger we put people in, which gave her an idea: goats! She knew that goats eat just about anything, so she and a friend started a research project to see if goats could reduce woody fuel buildup in our forests. In the process, she discovered that goats also ate a wide variety of troublesome weed species. She figured this information would be useful to ranchers, so she began to tell them to add five goats for every cow in order to improve their pastures.

"They just looked at me like I was insane," Voth said in an interview in the online magazine *On Pasture*. "Most ranchers don't want to have goats because they require a completely different kind of fencing and the market is much more difficult to access than the beef market. These were very good reasons and they made sense to me. But I'm not the kind of person you can just say no to."

So she turned her attention to cattle.

She also turned to animal scientists at Utah State University, who knew that a food's palatability is heavily conditioned by experience. When an animal finds a food that meets its nutritional needs it will choose this food over and over. That's because foods that taste good generally have more nutrients than toxins. Nutrients send positive signals to the brain. Conversely, toxins send negative signals, such as nausea, causing us to avoid foods that "taste bad." Flavor, in other words, is the brain's way of screening nutrients from toxins. (Of course, too much of a good thing can be toxic as well.)

According to Voth, this balance makes weeds ideal forage for cows because many invasives are high in nutrients and low in toxins. In fact, most weeds are at least as nutritious as grass and often higher in protein.

"That means if we can get a cow to try a weed, she'll continue eating it year after year," Voth wrote in an essay in *On Pasture*. "As a bonus, she'll gain weight at rates expected for an animal eating a higher protein diet. Thus, not only do we eliminate the cost of herbicides, we increase profits due to increased weight gain."

But how do you get a cow to like a food it has never eaten before? Because inexperienced animals are more likely to try new things, Voth focuses on young cows and gives them a lot of positive experiences. Here's how it works.

Invasive weeds are a big problem in the US, and livestock can be an effective alternative to poisons in controlling them. *Photo courtesy of Kathy Voth*

For the first four days of the training period Voth feeds the cattle unfamiliar but tasty (nutritious) food in tubs twice a day, including beet pellets, wheat bran, and hay cubes. Soon the animals associate her arrival in the pasture with a tasty meal. Combined with the natural competiveness of animals at feeding time, this means that they will try almost anything. On the fifth day, Voth serves weeds with some of the feed the trainees have already tried. She repeats this for three more days, increasing the amount of weeds and reducing the other foodstuffs until the mix is 100 percent weeds. Eventually, if the weeds are recognized in a pasture, the cattle will start eating them—and keep eating them ever afterwards (because they are yummy). Soon the trainees are training other animals. Voth has seen 12 cows train 120 more!

As a bonus, newly educated cattle are open-minded to trying other weeds in a pasture, even if they haven't been trained to eat them.

According to Voth, cattle will eat the following weeds: Russian, Canadian, Italian, Scotch, and musk thistle; diffuse, spotted, and Russian knapweed; yellow and Dalmatian toadflax; white top/hoary cress; leafy spurge; goldenrod; fringed sage; field bindweed; yellow and purple starthistle; wild licorice; horehound; common mullein; rabbit-brush; and many others. Voth has even trained cows to chow down on brush, including wild rose, willow, ash, and even mesquite.

Weed thorns and spines don't bother cattle. Voth has seen them eat cactus. As for toxins, her advice is to make sure the weed is safe before

Cattle can consume invasive cactus species—the thorns don't bother them! *Photo courtesy of Kathy Voth*

you start. (She keeps an updated and comprehensive list on her website.)

Then there are the economic benefits, such as not spraying herbicide. "Say you're a typical Western rancher and you have 400–500 cows," she wrote in *On Pasture*. "You train 50 of them and within a year they'll have trained all the rest. The cost of training those fifty cows is about $250 and you'll never have to do it again. On average, a gallon of herbicide costs $250 and it will treat not nearly as many acres as the cows will. It just makes sense to me."

Practicalities aside, there is also an important philosophical lesson to Voth's innovative strategy: real progress happens when we work *with* nature's principles instead of against them. Voth expressed this attitude well in a haiku she wrote for a column titled *The Tao of Cow*:

> *The war on weeds ends*
> *When cows begin to eat them*
> *Foe becomes forage*

Biological solutions are often right in front of our eyes—if we allow ourselves to look!

TO LEARN MORE

For more information about teaching cattle to eat weeds, visit Kathy Voth's website: www.livestockforlandscapes.com and the online newspaper *On Pasture*: www.onpasture.com

Goats are an option as well. For more information, see: www.goatseatweeds.com

The Low-Stress Way

Treating animals with kindness and respect

How we treat animals not only speaks volumes about who we are as human beings, it says a lot about how we treat the natural world generally.

When I first heard about a type of livestock handling that emphasizes patience and kindness toward animals, I thought "That's nice, but what has cow whispering got to do with sustainability?" At the time, I was interested in practices that improved the health of soils, plants, creeks, and people. Becoming buddies with cattle seemed at best a luxury and at worst an unnecessary expenditure of time and effort. Besides, I could easily imagine how most cowboys felt about the idea.

That was precisely the point, as I learned later during a low-stress herding clinic taught by Guy Glosson, Tim McGaffic, and Steve Allen. McGaffic opened the class with a history lesson about how livestock have been cussed, prodded, and badly manhandled for generations. Stressing cattle out, especially at roundup or during transport, was part of ranching culture and it's still standard practice on many ranches today. According to the clinic's instructors, however, high-stress handling of animals is wrong both ethically and economically. Stop the whooping and hollering when moving cattle, they said. Throw away your electric prods, wooden sticks, and aggressive attitudes. Conventional ideas of controlling animals by use of fear, pain, or other forms of stress-inducing pressure are counterproductive.

"Consider not wearing sunglasses when approaching cattle," said Glosson, who manages a ranch in west-central Texas. "You're the predator and they're the prey, or at least that's the way they look at it. If they can't see your eyes, it may make them more nervous, as they may not be able to judge your intentions."

And understanding the intentions of human beings is critical to the successful movement and placement of livestock, especially in the planned grazing systems that are an important part of sustainable ranching practices, because they require frequent moving of herds of livestock. Cooperation is always preferable to confrontation.

"If cattle get worried," continued Glosson in his soft drawl, "you've taken the first step toward losing control of the herd. Animals want to feel secure. But they won't feel secure if you're yelling at them all the time. Your job is to treat them with respect."

That's *not* how we usually behave toward cattle. Perhaps that explains why Baxter Black, a well-known cowboy poet and former large-animal veterinarian, once challenged Glosson over the idea of low-stress handling with a simple, steely, "*Why?*"

"I told him that it's all about the health of the animal," Glosson said. "Consistently handling animals without scaring them allows trust to be formed. This trust helps animals remain calm, and that equates into a healthier immune system and better response to vaccines and other medications they may need."

"I also told him that it was less stress on the handler, too, which made us healthier," said Glosson, with his easy laugh. "But I don't think I convinced him."

Another reason to adopt the low-stress way is economics. The margin of profit on livestock for ranchers is literally counted in pennies per pound. The stress put on cattle as they move from the ranch or the slaughtering facility can "shrink" an animal's weight as much as 15 percent. Stress can make an animal more susceptible to disease, often requiring additional medicines and additional costs. It can also affect pregnancy rates in cattle, the bread-and-butter of a rancher's bottom line. It all adds up quickly in dollars and cents.

Rancher Guy Glosson teaching a clinic on how to handle animals with kindness and respect. *Photo courtesy of the Quivira Coalition*

"For grazing animals like cattle, the most dangerous predator on Earth is a young human male," Glosson said. "Until trust is established, animals will always perceive humans as a threat. And we don't want that. These animals are now domesticated and for the most part they depend on people for their every need. If we want them to perform at their best, they must not be afraid of the person caring for them."

The low-stress way starts with recognizing the predator-prey relationship and the effects of such things as noise, size, distance, and motion on cattle, which like many animals have well-defined zones in which particular actions trigger particular responses. For example, the *recognition* zone is where the animal takes notice of you and tries to determine your intent. The *flight* zone, when crossed, will cause the animal to move away from your approach. Suddenly violating this zone means that you are likely to encounter an angry or panicked animal who has perceived you to be a threat.

According to Glosson, the key to successful low-stress handling is called "pressure and release." Your presence (as predator) creates pressure that an animal (as prey) wants to relieve. The critical moment occurs when you reduce the pressure instead of allowing the animal to do it for you by fleeing. You accomplish this by stepping into the animal's flight zone in such a way as to pressure it in a direction or manner that you intend for it to move, and then backing off when the pressure is no longer needed—before the animal runs away from you. Worked this way, an animal learns from the release of pressure, not the pressure itself, and a mutual understanding is established.

The whole idea is to use a law of nature to positive effect—herding instincts in this case. For example, Glosson teaches his students to

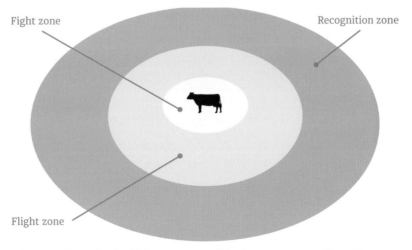

Diagram of an animal's flight zone, part of the low-stress way of handling livestock.

approach animals on foot in a nonthreatening manner, often zigzagging as they get closer. When an animal sends a signal, such as raising its head or widening its eyes, the student stops or backs up. If the animal moves off, then the student is too close or has done something threatening. Glosson tells them to start over.

"You're trying to start a conversation with the animal," said Glosson. "You're not trying to tell him you're a nice guy or anything, because you're not. You're still the predator. Instead, you're trying to communicate mutual respect. And you want to keep the conversation going as long as necessary to get the job done."

The originator of the modern concept of low-stress management, and Glosson's mentor, was Bud Williams, a Canadian rancher who spent his entire life studying how to handle animals respectfully and easily, including reindeer, elk, sheep, and wild cattle. The key, according to Williams, was to pay attention to the instincts of the animal. "We need people that are more sensitive to what the animal is asking us to do," Williams told an interviewer. "If we would be more sensitive to that, then these jobs that we work on would be so much easier to do."

And in the bigger picture, to support a system that produces healthy food in a regenerative manner and also sequesters carbon in the soil, we all need to be sensitive to what animals—and nature, by extension—are trying to tell us. It's all about communication, not just between humans and animals, but between people too. If you can't communicate clearly what you're doing, you probably also can't get it accomplished. And if you use high-stress methods, then there's a good chance what you accomplish won't last.

"We always work at a level where we barely get it done," Williams said. "We get as good as we need to get. We've reached a point now where we need to get better."

That point means treating the world in a manner that we would like to be treated ourselves—with kindness and respect.

TO LEARN MORE

To learn more about the Bud Williams method of livestock handling, check the information and instructional videos available at www.stockmanship.com

A good overview of low-stress handling techniques is available at Dr. Temple Grandin's website: www.grandin.com/B.Williams.html

Herding Wisdom

A scientific twist to an ancient art

Here's an essential equation for meeting the rising social and environmental challenges of our era: art + science = wisdom.

It's been a fruitful equation over a long period of time, but these days art and science spend more time in isolation than comingling, it seems to me, with the unhappy consequence of a general decline in wisdom among societies. Fortunately, a small but growing number of scientists and regenerative practitioners of traditional agricultural arts are finding new ways to work together. For example, two agricultural researchers, Fred Provenza, emeritus professor of behavioral ecology at Utah State University, and Michel Meuret, an animal ecologist at the French National Institute for Agricultural Research (INRA), have come up with the innovative idea of engaging skilled livestock herders as ecological doctors to heal damaged land and provide ecosystem services to human communities, including high-quality, locally sourced food and fiber—to which I would add increased carbon sequestration in soils.

Meuret and Provenza explain their thinking in a book they've edited titled *The Art and Science of Shepherding: Tapping the Wisdom of French Herders.*

Herding refers to the process of moving a group of animals as a unit from place to place in search of fresh forage under the daily direction of one or more humans and their ever-watchful dogs. The key to successful herding is managing the natural behaviors of wild or domestic herbivores, which historically lived in extended families for protection, and directing their movement and placement for specific purposes by human voice command and the swift movement of trained herding dogs.

The art of livestock shepherding is ancient. Humans have been moving cattle, sheep, and goats across varied landscapes at least since these animals were first domesticated in the Middle East, roughly eight

This shepherd in the French Alps uses a combination of art and science in directing his flock. *Photo courtesy of Michel Meuret (INRA)*

thousand years ago. Herding remained a widespread and ever-evolving practice around the globe, fitted to local conditions, until petroleum and barbed wire fencing put most shepherds out of business, including those in northern New Mexico, where I live. Today herding is vibrant mostly where it is essential to the cultural identity of indigenous people, such as the Masai of eastern Africa and the Navajo of the American Southwest, or where it has stubbornly persisted in the face of rampant industrialization, as it has in France, Italy, and Spain.

So what, you might ask? If it's a matter of controlling livestock, why not just use fences, as many progressive managers do? Isn't hiring a human shepherd more expensive and potentially problematic? If it's the positive effects of grazing animals on the land you seek, such as the removal of old grass and the fertilization of plants by animal poop and pee, shouldn't multipaddock, planned grazing strategies using portable electric fencing be sufficient? Alternatively, there's the high-tech control of livestock using GPS satellites and a radio collar, which gives the animal a mild electric shock when it strays off course. Why consider something as old-fashioned and low-tech as animal husbandry?

For Meuret and Provenza, based on their long-term research in France and America, the answer is simple: herding works.

"Even with a herder's salary," they write in the introduction to *The Art and Science of Shepherding*, "skilled herding is a low-cost way to address ecological, economic and social challenges and opportunities for improving the vigor of soil, the biodiversity of plants for the health of domestic and wild herbivores, and the health of people who rely on plants and herbivores for their well-being."

Meuret and Provenza believe that we've let fences, electric or otherwise, have too much influence on the foraging behavior of livestock and thus on the health of soil and plants. Fences can't do what a knowledgeable herder can do to optimize grazing from a diversity of forage resources over days and weeks (even hours). They see herding as essential to ecological doctoring because so many rangelands across the planet exist in a degraded condition from historical overuse. Grazing, they insist, when skillfully applied can do wonders to restore these rangelands to health, and livestock herding can be a big part of the solution.

The great conservationist Aldo Leopold understood this idea when he wrote in *Game Management* (1933) that wildlife habitat could be restored through the "creative use of the same tools that have heretofore destroyed it—axe, plow, cow, fire, and gun." It's the purpose

Using voice commands and a herding dog, a shepherd can carefully control the placement of a flock to accomplish nutritional and ecological goals. *Photo courtesy of Patrick Fabre/Maison de la Transhumance à Salon-de-Provence*

for which a tool is used that matters, not the tool itself (though we might drop the plow from the list today). If the goal is habitat restoration, then a whole set of tools is available—lying around in the old toolshed!

Herding is one such tool. By carefully designing and adjusting daily grazing circuits, a skilled shepherd can stimulate the appetites of his or her livestock by encouraging the herd to eat from a mix of plants, some highly palatable and others less palatable. This not only improves the nutrition, health, welfare, and production of the animals, it can be designed for specific ecological effects as well, such as chomping on weeds or targeting a particular species of plant.

Take sheep. According to Meuret and Provenza, there is a three-step process by which a flock is taught to respect grazing boundaries (that is, how far the sheep are allowed to graze):

1. Upon entering a new grazing sector in a pasture, the shepherd lets the flock approach a predetermined boundary. He or she stays in a visible place on the front side of the flock and also places the herding dog in motion along the boundary. When the first sheep come near the boundary line, the shepherd loudly shouts something like "Hôôô!" From prior experience, the flock knows this cry means that the shepherd disagrees. The sheep quickly change direction (while keeping eyes on the dog).

2. The next day, when grazing the same sector, the shepherd places the dog on the boundary again, but this time tells it to remain motionless. It's just a reminder for the flock, which usually turns of its own accord when approaching the boundary. However (sheep being sheep), if a part of the flock insists on crossing the boundary, the shepherd will shout again, from the same spot, causing the sheep to turn back.

3. During subsequent days, if the flock tries again to have a peek over the forbidden boundary, then the shepherd will cry again, though this time from behind the flock. This completes the training process; the flock now understands that this movement is totally off limits.

In this way, a shepherd directs the foraging behavior of the herd in precise ways to accomplish whatever goal he or she desires, whether it is nutritional or ecological. This is where the science comes in. Decades of research into animal behavior, dietary needs, plant toxins, wildlife habitat management, fire risk reduction, biodiversity requirements, and many other fields of scientific endeavor can inform the daily goals and hourly choices made by shepherds. In turn, shepherds provide observations, data, and other forms of feedback to the researchers from their experience on the land—something else that fences can't do. Herders also help keep wolves and other predators at bay, which is a growing challenge in Europe.

Best of all, none of this is theoretical—shepherding is alive and well in France, thanks to four state-supported herding schools and a steady flow of curious young people (mostly from cities) willing to sign up. And what the students learn in these schools and in the Alps is *wisdom*: the nonstop, mutually respectful and reinforcing give-and-take between art and science.

Let the doctoring begin!

TO LEARN MORE

The Art and Science of Shepherding: Tapping the Wisdom of French Herders, edited by Michel Meuret and Fred Provenza. Acres USA Press, Austin, TX, 2014.

"Shepherds' Know-How Faced with Globalization and Nature Conservation: A French Experience" by Michel Meuret and Mick Gascoin. *Quivira Coalition Journal*, no. 32, April, 2008. www.quiviracoalition.org

The Flerd

The many benefits of coexistence on the range

If we can break out of our belief silos somehow, all sorts of good things can happen.

Take sheep and cattle, for instance. Almost everywhere on the planet, they are managed separately from one another, the result of a deeply held belief in agriculture that the two herbivores have incompatible needs and behaviors. Not only that, most sheepmen and cattlemen consider *each other* to be incompatible as well—even to the point of physical violence during various nineteenth-century range wars. Don't try to convince Eric Harvey of this incompatibility, however. He credits the explosion of native grass on his farm in New South Wales, Australia, from 7 to 130 species in only 7 years, to planned grazing and a combined flock of sheep and a herd of cattle—called a "flerd."

I had the good fortune to witness the success of this coexistence in person.

The story begins in 2004, when Harvey, a gregarious former wool trader, purchased the 7,000-acre farm, called Gilgai, located a few miles from the crossroads city of Dubbo. A few months later, however, Harvey nearly "bought the farm" himself when he had a massive heart attack. After recovering, he was astonished to learn from his doctor that his body was almost completely devoid of minerals essential to human health. He knew there weren't many minerals in his water supply—due to groundwater scarcity, Australians collect and drink a lot of rainwater—but he assumed that he was getting enough minerals from the plants and meat that he ate, which in turn get their minerals from the soil. However, 95 tests showed he wasn't.

This news was a huge eye-opener for him, he told me.

Harvey had soil tests conducted at Gilgai and discovered that the land was also depleted of essential minerals, including carbon. This

Australian flerd farmer Eric Harvey stands beside a rainwater collection tank on his farm. Harvey recovered his health in part due to the improved conditions of his land after he began using innovative grazing techniques. *Photo by Courtney White*

meant that the farm and his health were now one and the same—both had to recover somehow. But where were the minerals going to come from, he wondered? A mine? A factory? That didn't sound very practical or economical. And what about carbon? He knew there was a lot of organic material in compost, but was he supposed to spread it over all 7,000 acres of land? That didn't sound economical either.

A chance conversation with a neighbor provided him with an unexpected answer: the sky. Carbon was freely available in the air, his neighbor said, in the form of carbon dioxide, and all Harvey had to do was get it into the soil via photosynthesis, livestock, and planned grazing practices. The goal, he told Harvey, was to grow native grass—diverse and copious amounts of it.

So that's what Harvey did. First, he studied the principles of planned grazing, originally pioneered by Allan Savory. Then, after deciding to put them to work, he made another unconventional decision: to run cattle and sheep together as one grazing unit. Years earlier he had seen sheep and cattle grazing together on a farm in Africa and thought "that makes sense," even though he knew that conventional thinking ran exactly in the opposite direction. Harvey ignored all that, however, and in 2005 he created his first flerd, eventually comingling five thousand sheep and six hundred cattle.

His goal was to use the different grazing behaviors of sheep and cattle to benefit plant vigor, diversity, and density. Mixed-species grazing happens all the time in nature, Harvey said. Herbivores complement each other in what they eat, the composition of their manure, and the way they move across a landscape. As Harvey described it, grazing animals create an organic "pulse" of energy *below* the ground by their activity, meaning that as plant roots expand and contract with grazing, carbon is fed to hungry microbes, fungi, protozoa, and nematodes, which in turn feed minerals to the plants. Round and round. The animal manure above ground provides additional fertilizer, which helps with nutrient cycling. Harvey said his plan for the flerd was to make the belowground "pulse" beat as strongly as possible.

To accomplish this goal, Harvey divided the 7,000-acre farm into 196 paddocks, mostly with electric fencing, creating an average paddock size of 140 acres (the smallest is 6 acres). The flerd moves from paddock to paddock every few days, giving each paddock plenty of time to grow more grass. And with only one "mob" to watch, Harvey is often back home by 10 a.m. As further work reduction, Harvey pays for a service that provides aerial images of his farm daily, which allows him to monitor the growth rate in his paddocks at a tiny scale. He calls this service "pastures from space" and says it gives him an invaluable snapshot of forage conditions, which helps adjust his grazing schedule.

Harvey also ground-truths the monitoring data he receives. That's how he knows he has been able to expand the number of plant species on Gilgai from 7 to 130. This improvement in diversity has substantially enhanced the mineral content of the plants, thanks to better functioning water cycling and carbon cycling in the soil, as well as access to minerals deeper in the soil profile due to increased root growth. And when these plants are eaten by animals, which are in turn eaten by us, the minerals enter our bodies, as Harvey can personally attest—his physical health has improved dramatically.

Which is why Harvey grows and sells only grassfed products from the farm.

As for the flerd itself, Harvey has hardly any trouble running sheep and cattle together. The key is to raise them as one family, he said,

A flock of sheep and a herd of cattle grazing together is called a flerd. On Gilgai Farm, the flerd has helped the land regain biological health. *Photo courtesy of Gilgai Farms*

especially the lambs. Sheep will bond with cows at a young age and remain bonded for the rest of their lives. As a result, the sheep follow the cattle wherever they go, which means they'll move from paddock to paddock with the herd without much fuss, which is great news for a multipaddock farm like Gilgai. It also means Eric doesn't have to train any sheep to electric fencing, only the cattle.

"Needless to say, moving one herd of livestock is a lot easier than moving two," he said. "You just have to make sure there's enough forage and water ahead of them."

The only real trouble he's encountered happens during calving, when mama cows become highly protective and might kill an ewe that comes too close. Eric solves this by separating the cattle from the sheep during their respective birthing seasons. "The only other conflict I've ever seen is over shade," says Eric. "And that's been minor. Otherwise, they get along great."

Another benefit to a flerd is protection from predators, such as coyotes and wolves (which don't exist in Australia). Experiments have shown that when sheep are bonded to cattle they are protected from predators, which are reluctant to take their chances with a closely packed herd of bovines. Experiments also demonstrated that sheep gain weight faster when grouped with cattle compared with sheep managed as a separate flock. Wool production was also greater with the flerd than with sheep foraging alone—a fact that Harvey said he could confirm. He attributed both improvements to the healthier soil and increased diversity of plants on Gilgai.

"With all these benefits," I asked, "why don't more farmers and ranchers give flerds a try?"

"It's a paradigm thing with humans," he responded. "It's not an issue in nature."

That means it's time to break out of our silos!

TO LEARN MORE

For more about Eric Harvey and Gilgai Farms,
visit: www.gilgaifarms.com.au

"Flerds: Sheep and Cattle Grazing Together for Predator
Control and Pasture Management" by Kathy Voth.
On Pasture, June 23, 2014. www.onpasture.com

Animal Power

The comeback of a regenerative source of energy

S ometimes moving forward requires going backward.

One ancient practice nearly wiped out by the march of "progress" was the widespread use of animal power in many important endeavors, including farming, logging, and various types of transportation. In the late nineteenth century, for example, getting around New York City meant employing one or more of the nearly 200 thousand horses stabled in the city (whose manure production posed a serious and perennial public health hazard). Equally hard to imagine today is that until the adoption of tractors in the 1920s, nearly all American agriculture was powered by livestock.

As someone who came of age among the asphalt suburbs of Phoenix, Arizona, during the 1970s—the nadir years for animal power in the US—it was hard for me to comprehend these historical facts when I first heard them. Although I had spent my youth around horses, they were strictly the recreational variety. I knew nothing about draft animals or horse farming, except that they had become anachronisms, replaced forever by petroleum, or so I assumed. Therefore, it came as a surprise in the 1990s to learn that animal power was making a comeback—draft horses in particular—propelled by rising concerns about carbon pollution and oil scarcity. But what exactly *was* animal power?

In 2008, I decided to find out. In early July, I traveled to the heart of Amish country in central Ohio to attend an annual event called Horse Progress Days, which is partly a celebration of horse farming and partly a convention of farmers gathering to witness the latest in animal-powered "technology"—a word that must be used judiciously given the famous Amish disdain for gadgetry. The most educational part of my trip, however, happened on the evening of my arrival.

An Amish boy at Horse Progress Days in Ohio, an annual event that celebrates animal power and showcases the latest developments in horse farming. *Photo by Courtney White*

Standing at the second-story railing of my hotel, I watched an Amish family bale and load hay in an adjacent field. The hay had been cut a day or two earlier to dry and now needed to be put up before the leaden sky began to drizzle. There was a calm, methodical urgency to the family's work. The apparent patriarch, wearing the standard Amish uniform of straw hat, plain shirt, suspenders, and black pants, stood in a hay baler that was so old it looked like it belonged in a history museum. It sounded old, too. Its single-stroke engine, whose job was to compress the loose hay into a square bale and bind it with string, sputtered and choked so noisily that I expected it to give up and die at any moment.

The baler kept going, however, pulled by a team of handsome black draft horses that I later learned were Percherons. Together they spiraled steadily toward the center of the field, the baler excreting—for that's what it looked like—a tidy green bale of hay every 30 seconds or so. Not far behind followed another team of horses, guided by a young Amish man, likely a son or son-in-law, who stood on a flatbed wagon. On the ground were three young women, in plain dresses and white bonnets, who loaded the wagon with the freshly minted bales. The work must have been pleasurable because I heard the sounds of laughter from where I stood. When they filled the wagon, the youngsters drove it to a farm across the busy road, returning a short while later to continue their rounds.

In less than an hour, their work was done. Completely emptied of hay, the field looked like a shorn sheep, bewildered and turned back to pasture. I was sort of bewildered too. *That didn't look so hard to do*, I thought. My mood changed to astonishment a short time later, however, when I heard the sound of another engine fire up. This was not the sound of a coughing relic, however; it had the confident hum of serious machinery.

Indeed, it belonged to a John Deere combine of some sort (I knew as much about farm machinery back then as I did about draft horses). Within a minute or two it began sweeping across a neighboring hay field of approximately the same size as the Amish field, chased almost comically by a tractor pulling a large bin on wheels. The combine sucked up the loose hay from the ground and then spit it—for that's what it looked like—through a long pipe into the careening bin beside it. Idling nearby, with their lights on and engines running, were three more tractors with bins, waiting patiently for their turn.

In about half the time it took the Amish family to bale and load their hay, the combine had finished its work. All four bins had been filled and the tractors dutifully dispatched someplace over the horizon with their green cargo. The combine, too, took off down the road for parts unknown.

And suddenly all was quiet.

What had just happened? Two fields of similar size had just been cleared of hay—one principally by horses, the other by horsepower. I wondered: How many gallons of precious diesel had the ancient, coughing baler used in comparison to the purring combine and speedy tractors? The difference must have been huge. And where did all that industrially gathered hay go? How many miles down the road would it travel to its ultimate destination? I had no idea, but I knew exactly where

A demonstration of horse farming at the county fairgrounds in Madras, Oregon, organized by the *Small Farmer's Journal*. *Photo by Courtney White*

the Amish hay went—across the road, to be used, I'm sure, to feed the farm's dairy cows in the coming winter. The contrasting images bounced around in my mind as I soaked up the silence.

Years later, the contrast has only sharpened.

Animal power, of course, isn't just for the Amish. It's being implemented across the nation, especially by young people. Concerns about carbon pollution and our dependence on petroleum have only grown since I made my journey to Ohio in 2008, making draft horses, oxen, and other livestock increasingly attractive power sources—and I mean that literally. At the Horse Progress event, I witnessed a Belgian horse walking steadily on a treadmill that turned a small electricity generator, which powered a number of household appliances. It gave new meaning to the term *horsepower*!

There are pros and cons to using animals in agriculture, of course, including the high cost and long hours involved in keeping animals. Mistreatment is a concern as well. And there's the philosophical argument that putting livestock "under the yoke" is a form of animal abuse. I suspect that the final decision may come down to whether you are an animal person at heart or not and whether you love your animals and treat them with respect. I love horses, so the appeal for me is direct. But I'm also aware of the complexities and challenges involved with draft animals. That's why some people will choose to go the solar or wind or biodiesel route for their energy needs, which is understandable. For those who would rather rely on animal power, however, it's good to know this option not only exists but is making a comeback.

In fact, I'll bet that draft animals will be part of any regenerative system we develop to meet the expanding challenges of the twenty-first century!

TO LEARN MORE

The *Small Farmer's Journal* is a great resource
for horse farming: www.smallfarmersjournal.com

"Draft Animal Power for Farming"
by Tracy Mumma (2009) is available through
the ATTRA website: www.attra.ncat.org

The Mobile Matanza

Slaughterhouses—fixing the weak link

W hen implementing regenerative solutions, it is important to remember that a sturdy chain is only as strong as its weakest link.

A critical problem for many small-scale ranchers, local food advocates, and rural development entrepreneurs is the chronic lack of nearby slaughterhouses. That may sound prosaic, but too often producers must drive long distances to have their animals processed, which raises their costs and reduces their profitability—and thus jeopardizes the sustainability of their operations. It's a weak link in many rural communities, but one that has fostered an innovative solution: instead of driving to the slaughterhouse, have it come to you!

I was introduced to this idea on a sunny day some years ago, when I drove to Taos, New Mexico, to attend a ribbon-cutting ceremony for a mobile slaughtering unit (MSU), only the second one in the nation at the time. I joined a sizeable crowd of local ranchers, farmers, and others at the headquarters of the Taos County Economic Development Corporation (TCEDC), a nonprofit codirected for over 25 years by Pati Martinson and Terrie Bad Hand. Both women are Native Americans and have extensive backgrounds in social justice work, education,

Taos County Economic Development Corporation's codirectors Terrie Bad Hand (left) and Pati Martinson at a debut ceremony for the mobile slaughterhouse. *Photo by Courtney White*

economic development, youth leadership, and civic engagement. They dubbed their MSU a *matanza*, a Spanish word for a communal celebration involving the annual harvest of a farm's livestock, an old tradition in the region.

The definition of the mobile matanza is simple: it's a semitruck carrying a self-contained slaughtering lab and cold-storage unit that drives to a farm or ranch and processes the livestock on-site under the watchful eye of a professional inspector (usually from the USDA). The truck then takes the carcasses to a local cold-storage facility, where they hang for 14 to 21 days before being cut, packaged, labeled, and frozen for sale or further storage.

The idea of a mobile unit originated among a ranchers' cooperative in the San Juan Islands in Puget Sound, near Seattle. Isolated on their respective islands, the ranchers were frustrated with the high cost and logistical difficulty of taking their animals to a slaughterhouse on the mainland, so they decided to build a ferry-friendly facility on wheels instead. In 2002, after a lengthy period of trial and error, they arrived at a design that met their needs and those of various regulators, and the nation's first MSU went to work.

The idea came to New Mexico in 2006, when the state legislature, backed by Governor Bill Richardson, approved a $200,000 appropriation to purchase a mobile unit and entrusted the project to TCEDC. Shortly thereafter, Martinson and Bad Hand made a field trip to Puget Sound to see how an MSU operated. They were struck by its potential for northern New Mexico.

"The isolated islands we saw were like the isolated villages around Taos," Bad Hand told me. "It worked there and we thought it could work here."

It has—and well—as I observed during a field visit of my own. Back in 2006, however, the mobile matanza faced a series of intimidating hurdles before becoming fully operational.

Meat inspection. The USDA wasn't interested initially in the TCEDC project, and the state meat inspection system had just been suspended by the governor.

Bias. It was necessary to overcome long-standing prejudices by agencies and regulators against local, family-scale producers, which Martinson and Bad Hand viewed as a civil rights issue.

Adaptation. Built in Washington, the MSU had to be adjusted to New Mexican conditions (such as higher ground clearance) and different needs (such as bison slaughter)—not to mention driving the unit all the way home to Taos.

Mobile matanza operator Gilbert Suazo Jr. on a job site. A USDA inspector travels with the slaughterhouse wherever it goes. *Photo by Courtney White*

Money. The New Mexico legislature provided zero funds for staffing the MSU, maintaining it, or advertising its services.

Job descriptions. None existed, so they had to be made up from scratch.

Cut-and-wrap facilities. Where was this vital service going to be done? All the local options around Taos were going out of business or had scaled back substantially in recent years, leaving an important hole in the business plan.

Regulators. Before the matanza could open for business (and stay in business), nine different regulating authorities had to sign off, including organic certification, transportation, the state Environment Department, weights and measures licensing, the Livestock Board, the USDA, and even Homeland Security.

To meet these challenges, Martinson and Bad Hand had to be creative. On the cut-and-wrap front, for example, they went to the largest refrigeration company in the nation, Polar King, and asked them to custom-build an 800-square-foot, self-contained fiberglass facility for cutting, hanging, and wrapping, to be housed at TCEDC. "They were great to work with," Martinson said, "and we were able to install it very quickly."

In addition to being mobile, the matanza is highly humane, as I witnessed during my visit. A USDA inspector is on-site to ensure that animals are healthy, well treated, and feel as little pain as possible when they are killed. The whole process is held to the highest standards, the inspector told me.

The matanza can process hogs, cattle, bison, sheep, goats, captive elk, and yak. Output can average four to ten animals a day, depending on size. Its operational range is a one-hundred-mile radius of Taos (with a few exceptions), and its regular customers include tribes, local ranchers, and restaurants. TCEDC charges a fee for the slaughter and for the cut-and-wrap.

The start-up cost of the project was under $500,000—$100,000 each for the slaughter lab, the semitruck, and the cut-and-wrap facility, plus labor and overhead. That's *only* $500,000, by the way—not much money for all the benefits. For example, Bad Hand said that the matanza processes about 100 thousand pounds of meat each year, meat that is produced by the community and stays in the community, as does the money it generates. This includes the added value it brings to the ranchers via sales at farmers markets and at grassfed premium prices, which in turn has a positive impact on maintaining the land-based traditions and culture of the area. The Food Center at TCEDC assists ranchers with labeling and marketing too, if they want it.

Of course, the primary limitation of an MSU is its capacity. While it's ideal for family-scale producers, its inability to handle more than ten animals a day means it's probably not an option for medium-sized producers—many of whom continue to suffer from the lack of local slaughterhouses. One solution involves a fleet of MSUs making the rounds (especially if fueled by biodiesel).

Bringing the slaughterhouse to the producer saves on fuel and transportation costs, and there's no need to warehouse animals in feedlots. Plus, food stays in the community in which it was raised, and cultural traditions are reinforced. The key, said Martinson, is to make the MSU part of a community food system. "That way it is part of a larger effort to help people."

The mobile matanza has strengthened a weak link in the Taos area and demonstrates the value of examining the whole chain of any regenerative process.

TO LEARN MORE

Taos County Economic Development
Corporation website: www.tcedc.org

For more on mobile slaughter units, visit
these two websites: www.mobileslaughter.com and
www.smallfarms.wsu.edu/animals/processing.html

PART TWO
Farming

Organic No-Till

Farming like the Earth matters

I f we could do one thing for the planet, I'd vote for ditching the plow. For starters, turning soil over with a plow to grow food—a practice that goes back at least five thousand years—causes erosion and robs plants of critical nutrients, organic matter, and shade by removing crop residues on the soil surface. Worse, plowing—also called tilling—destroys the microbial universe underground by exposing beneficial protozoa, fungi, and other forms of life to the killing effects of sunlight, wind, and heat. The plow itself also tears delicate and essential mycorrhizal fungi to pieces. These microcritters are a key to soil fertility, which is why synthetic fertilizers are required in conventional systems to replace the fertility lost because of their mass slaughter. Lastly, plowing also releases large amounts of stored carbon into the atmosphere, adding to the planet's greenhouse gas problem. Repeated plowing eventually depletes soil of its carbon stocks.

Fortunately, there is an alternative to the plow—it's called no-till. On a modern, conventional farm, a tractor and a plow are required in order to turn over the soil and prepare it for seeding and fertilizing, a process the often requires *three* passes of the tractor over the field. In a no-till system, a farmer uses a mechanical seed drill pulled behind a tractor to plant directly into the soil, requiring only one pass. The drill makes a thin slice in the soil as it moves along, but nothing resembling the broad furrow created by a plow. The soil is not turned over and any growing plants or crop residue on the surface are left largely undisturbed, which is a great way to reduce erosion and keep soil cool and moist, especially during the hot summer months. These are all good reasons why no-till has grown in popularity with farmers around the world.

One of the major disadvantages of no-till, however, is its lack of weed control. When farmers don't plow, the weeds say "thank you very much" for all that undisturbed soil and grow vigorously. To kill weeds in a no-till system, many farmers apply chemical herbicides to their fields. Lots of them. They also spray pesticides to keep the bugs in check. Additionally, many no-till farmers use genetically modified seeds, often in combination with the synthetic herbicides.

All of this is *verboten* in an organic farming system, of course, which brings us to the Holy Grail of regenerative agriculture: organic no-till. It combines the best of both worlds—no plow and no chemicals. It operates on biology—plus the tractor and the seed drill. It's a major development that's just catching on and it came about as innovations so often do—by accident.

One day, Jeff Moyer, the longtime farm director at the Rodale Institute, an organic farming research and education center located in eastern Pennsylvania, noticed that as he drove in and out of a field on his tractor, the wheels had crushed and killed a plant called hairy vetch along the field's edges. Vetch is a winter-tolerant legume that organic farmers often plant as a cover crop in the fall to protect the soil surface until the cash crop, such as wheat or corn, can be planted in the spring.

Moyer realized that by "crimping" (crushing) the vetch plants with the tractor's wheels, he had caused them to die but hadn't detached the plants from the soil. This was important because, by remaining attached to their roots, the dead vetch became a new type of cover crop—albeit a dead one. (Normally cover crops are harvested, composted, and applied to the field later in the year.) This was good because it meant that after a pass of the no-till drill in the spring—to plant the seeds of the cash crop—the layer of dead vetch would *suppress any weed that tried to grow*. Most weeds do not have the strength or stamina to push their way up through a layer of organic material. No chemicals needed; no turning the soil needed. Voila, organic no-till!

However, no mechanical piece of equipment existed that could do the job of crimping the cover crop (and driving a pickup truck through a whole field for the task would be impractical). So, Moyer took the initiative and, after lots of trial and error, he and his colleague John Brubaker settled on a design for what they call a roller-crimper—a hollow metal cylinder to which shallow metal ribs have been welded in a chevron design (like tractor tires). The roller-crimper is mounted in front of a tractor, and as it rolls along through a field it efficiently crimps the cover crop, breaking the plants' stalks. The weight of the roller-crimper can be adjusted by adding or removing water from the hollow cylinder.

As developed by Moyer and others over the subsequent years, there are four basic steps to organic no-till:

Detail of a roller-crimper, the key piece of equipment in an organic no-till farming system. *Photo by Courtney White*

1. To protect the soil and keep down the weeds, a winter-hardy cover crop such as vetch, barley, wheat, or rye is planted in the fall;
2. When the cover crop reaches maturity in the spring, the farmer knocks it down with a roller-crimper;
3. The farmer plants a cash crop with a no-till drill, usually at the same time she or he crimps (crimper in front of the tractor, drill pulled behind), and the cash crop germinates and grows up through the crimped cover crop;
4. After harvest in the fall, the organic residue of both crops can be incorporated into the soil as next year's cover crop is planted.

Organic no-till offers a wealth of benefits. The decomposing cover crop builds soil and substantially reduces erosion. Nearly all annual weeds are smothered. Cover crop roots increase nutrient cycling in the soil, and biodiversity is increased. Plus, greenhouse gas emissions are reduced. On the practical front, costs are low and the roller-crimper is easy to use and maintain.

Better yet, if the tractor runs on farm-produced biodiesel or if the crimper is pulled by horses, dependence on fossil fuels is further reduced.

There are, however, some downsides to the system. Planting cover crops is extra work and an extra cost, and choosing the correct cover crop for your land and matching it to the needs of the cash crop can involve a lot of experimentation. The crop requires water, sometimes a lot of it (which makes the practice problematic in arid environments). Perennial weeds can be a nuisance. Rolling the crimper too early in the

The roller-crimper in action (Jeff Moyer on tractor). A cover crop is crimped in the spring to smother weeds and protect the cash crop. *Photo courtesy of the Rodale Institute*

season can be a costly mistake if the cover crop doesn't die completely. And like anything new, success requires a great deal of patience.

Overall, the advantages far outweigh the downsides, which is why the practice is spreading rapidly. According to Moyer, there are now hundreds of roller-crimpers at work on farms and research stations across the nation.

Organic no-till farming has another big benefit: it sequesters carbon dioxide in the soil. Research at Rodale shows that when soil is turned over by a plow, the sudden access to oxygen speeds up the biological decomposition process, by which microbes eat up organic matter and "burp" carbon dioxide into the air. In contrast, organic methods sequester carbon by improving biological life in the soil. When combined with no-till, according to Rodale's data, the system has the potential to sequester 1,000 to 2,000 pounds of carbon per acre per year—which is a lot—pulled directly from the atmosphere.

Organic no-till is a Holy Grail that we can all appreciate!

TO LEARN MORE

Organic No-Till Farming: Advancing No-Till Agriculture—Crops, Soil, Equipment by Jeff Moyer. Acres USA, Austin, TX, 2011.

"Regenerative Organic Agriculture and Climate Change: A Down-to-Earth Solution to Global Warming," a white paper from the Rodale Institute, 2014. Available through www.rodaleinstitute.org

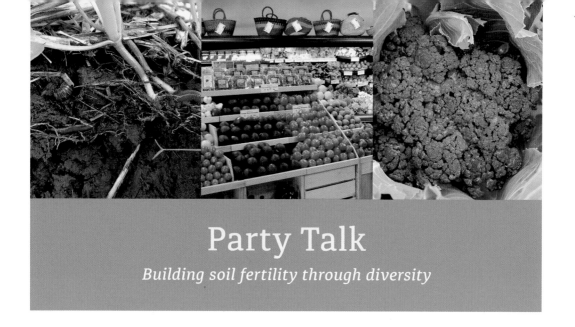

Party Talk
Building soil fertility through diversity

A spider hunt in a cornfield confirmed what I had suspected for years: diversity in plant, animal, and microbial communities is a key to their health and fertility.

The hunt happened at the end of a no-till, cover crop workshop on a scorching summer day in central Kansas led by soil scientist Jill Clapperton. Cover crops are planted by farmers at different times of the year in order to protect the soil surface with something green and growing. It's a great way to keep soil cool, moist, and stable, reducing wind and water erosion significantly. Most modern agricultural practices do not employ cover crops, however, preferring acres and acres of monocrops, such as industrial corn, during the growing months and bare soil during the winter. Nature, in contrast, likes cover crops (including weeds) and uses them extensively.

Soil profile in a cover crop field. Cover crops protect the ground from wind and water erosion while keeping the soil moist and cool. *Photo courtesy of Gail Fuller*

Nature also loves to talk. The best way to build soil fertility, Clapperton told the workshop participants, is to start a "conversation among plants." Cool-season (spring and fall) grasses such as barley, wheat, and oats, and cool-season broadleaf plants such as canola, peas, turnips, lentils, and mustard, need to dialogue constructively with warm-season (summer) grasses, including millet, corn, and sorghum, and warm broadleafs such as buckwheat, soybeans, sunflowers, and sugar beets. Who gets along with whom? Who grows when? Who helps whom?

If you can get these plants engaged in a robust conversation in one field, Clapperton said, you'll be creating "a feast for the soil." That's because increased plant diversity and year-round biological activity absorb more CO_2, which in turn increases the amount of carbon available to roots, which feeds the microbes, which builds soil—round and round. The complex community of life below ground involves a mind-boggling variety of relationships between plant roots, microcritters, minerals, water, and energy—a web of activity that sustains all terrestrial life on Earth. Encouraging this life to grow and be healthy creates all sorts of positive benefits for everything involved.

Take what happened on Gail Fuller's farm, which we visited during the workshop. When Fuller took over the operation from his father, they were growing just three cash crops: corn, wheat, and soybeans. Now Fuller plants as many as 53 different kinds of plants on the farm, mostly as cover crops, creating what Clapperton called a "cocktail" of legumes, grasses, and broadleaf plants. Fuller doesn't apply any pesticides and has dropped herbicide use by half, with a goal of nearly eliminating them altogether. Ditto with fertilizer as well. That's because he considers weeds to be a part of the dynamic conversation!

As a result of this robust conversation, Clapperton said, the carbon content of the soil on the Fuller Farm has doubled from 2 percent in 1993 (when they switched to no-till) to 6 percent as of 2012. That's huge.

There's more: the mineral content of Fuller's crops has also risen dramatically. This is important because all living creatures, humans included, need minerals and vitamins to stay strong and healthy. Iron, for example, is required for a host of processes vital to human health, including the production of red blood cells, the transportation of oxygen through our bodies, and the efficient functioning of our muscles. Copper is essential for the maintenance of our organs, for a healthy immune system, and to neutralize damaging free radicals in our blood. Calcium is essential for bone health. And every cell in the body requires magnesium to function properly.

A deficiency or imbalance of these "trace" minerals (so called because they are only needed in tiny amounts) can cause serious damage to our health, as most people understand. That's why taking vitamin pills has become such a big deal today—and big business—especially where young children are concerned. It's not simply because we don't eat our veggies or because we drink too much soda, it's because the

THE SOIL FOOD WEB

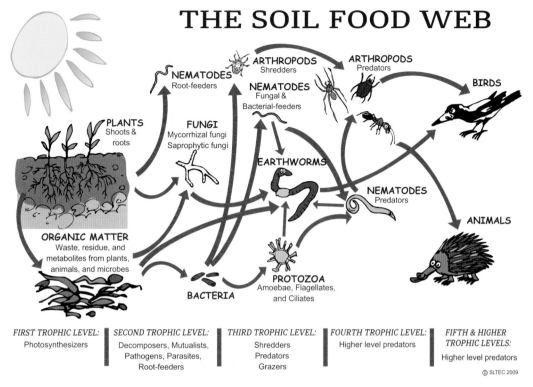

The soil food web sustains all life on the planet, from microbes to fungi to earthworms to mammals—with an intricate web of relationships. *Image courtesy of Sustainable Liquid Technology Ltd*

veggies themselves don't have the amount of essential nutrients that they once did. In some cases, the drop has been dramatic.

How did this happen? Well, industrial agriculture happened, but a specific culprit is what we've done to the soil. As a consequence of repeated plowing, fertilizing, and spraying, the top few feet of farm-land soil have been leached of their original minerals and stripped of the biological life that facilitates nutrient uptake in plants. According to the industrial mindset, as long as crops are harvestable, presentable, digestible, and profitable, it doesn't matter if their nutrition is up to par. If there's a deficiency, well, that's what the vitamin pills are for.

Gail Fuller reversed this trend in two ways. First, his use of no-pesticide, no-till farming keeps the microbial universe in his farm's soil intact and healthy. And when the soil dwellers have enough carbon (as an energy source), they facilitate the cycling of minerals in the soil. Earthworms, nature's great composters, are especially good at this. Second, a vigorous and diverse cover of crops puts down deep roots, enabling plants to access the fresher minerals in deeper soil, which then become available to everything up the food chain, including us.

Furthermore, by covering the soil surface with green plants or litter from the dead parts for as long as possible, Clapperton said, a farmer like Gail Fuller traps moisture underground where it becomes available for plants and animals (of the micro variety), enabling roots to tap resources and thus growing abundant life.

"Aboveground diversity is reflected in belowground diversity," Clapperton said. "However, soil organisms are competitive with plants' roots for carbon, so there must be enough for everybody."

The spider hunt at the end of the day drove home these points. Each participant was given a butterfly net and told to swish it 50 times in Fuller's corn field and then come back to show the workshop entomologist what we found. We found lots of spiders.

In a conventionally managed, monocropped Midwestern corn field, there would be no spiders, the entomologist told us. There wouldn't be much of anything living, in fact, except the destructive pests that could withstand the regimen of genetically modified seeds, industrially produced nitrogen fertilizer, and synthetic pesticides. In Fuller's field, however, the corn coexisted with a diversity of other plants in mutually supporting relationships. Weeds are good. Spiders are good. Diversity is great. This was a perfect example, Clapperton said, of why nature is such an important role model: to be healthy we need to recycle nutrients, encourage natural predators to manage pests, and increase plant densities to block weeds, all integrated and interconnected together. Good advice for humans too.

It all begins with a dynamic and diverse conversation at a cocktail party for plants—where everyone is gossiping about carbon!

TO LEARN MORE

View these YouTube videos about
soil health and cover crops:

Dr. Clapperton at the Quivira
Coalition's annual conference in 2012:
http://www.youtube.com/watch?v=o6daE2sYegg

Gail Fuller on Farming in Nature's Image:
www.youtube.com/watch?v=IRC8R2r9nJI

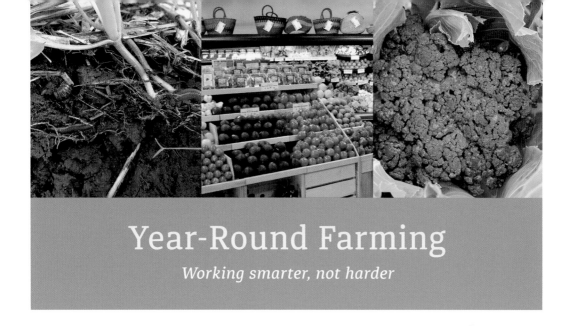

Year-Round Farming
Working smarter, not harder

K nowledge is an important key to providing stable, enjoyable, middle-class jobs producing healthy, nutritious local food.

That's certainly been the experience of farmers Paul and Elizabeth Kaiser, owners of Singing Frogs Farm, located near Sebastopol, in northern California. The Kaisers have developed a highly innovative model of organic no-till agriculture that sounds like it is more work, but is not. It's called year-round farming —as in no winter break. At all. "Every week we plant seeds and every week we harvest a crop, all year round," Paul Kaiser explained to me during a visit to the farm. "We didn't start that way, but it's the only way we'll farm now."

Kaiser calls it *knowledge-based farming* and credits scientific understanding of how nature works—especially the workings of soil microbiology— for the success of their model.

Here's how it works on their place: seeds are sown in a greenhouse, and seedlings are nurtured to a transplantable

Paul Kaiser and his wife, Elizabeth, have developed an innovative model of organic no-till farming that works year-round. *Photo by Courtney White*

When a crop plant is harvested on Singing Frogs Farm, a young plant takes its place, often within hours. *Photo by Courtney White*

age. When a crop is harvested in the field, a young plant takes its place, often within hours. This way, the farm never stops producing food—and does so without growing weeds or using cover crops and commercial fertilizers. Instead, the Kaisers produce a great deal of compost on-farm and spread it along each row of crops on top of the ground, rather than mixed into the soil as is normally done. The whole process is organic, sustainable, and profitable.

According to Kaiser, a typical organic farm in California grosses $12,000 per crop-acre per year. One of their organic farming neighbors grosses $20,000 per crop-acre, he said, which is really good. And Singing Frogs? It grosses over $100,000 per crop-acre per year! The Kaisers receive this return from the 120-member community-supported agriculture (CSA) program they operate and from the three farmers markets they attend each week—all within twelve miles of the farm.

Did I mention that Singing Frogs produces all of this food and money from only 2 acres of land? And the carbon content of the farm's soil has risen from 2 percent to 6 percent (measured annually at a depth of 12 inches) in just six years, to boot.

You're probably wondering, "How is it done? What's the catch?" It's all done by hand—and there isn't a catch. Yes, year-round farming is labor-intensive, Kaiser told me, but that's not a problem if you love what you do. Yes, labor costs are high on the farm, but that's because the Kaisers employ four people full time and pay them a decent wage (rather than rely on seasonal migrant labor). Providing stable, enjoyable, middle-class jobs with health benefits should be the goal of every farm, Kaiser believes. And as Singing Frogs demonstrates, it's not a pipe dream.

So how was this success achieved? "By a lot of thoughtful trial and error," he told me with a smile.

In 2007, the Kaisers spent five months looking for an organic farm to buy. They finally found one in a rolling landscape dotted with single-family homes intermingled with vineyards. It was a giant leap of faith—neither of them were farmers. Paul Kaiser has a background in agroforestry (with two advanced degrees) and Elizabeth Kaiser's background is in public health and nursing (she also has two degrees). They met in West Africa in the late 1990s, where Paul Kaiser worked for a nonprofit organization rehabilitating degraded farmland. Both wanted to farm, however, so with the support of their families, they decided to give it a go.

Being newbies, they chose the conventional organic model of production: grow one or two crops per year, employ tractors, and let the land go dormant in the winter. Two developments changed their mind after the initial season. First, lots of mechanical things kept breaking down, which frustrated them and cost a lot of money. Second, closing down the farm at the end of the growing season and reopening it again in the spring required an immense amount of energy on their part and required the use of heavy machinery (which kept breaking down). The Kaisers decided there had to be a better way.

The answer was to become a "knowledge-intensive" farm and work smarter, Paul Kaiser told me. This led to two radical changes during the second season of farming. One: replace the machines with people. Two: never stop farming.

Six years later, the Kaisers can say with confidence that it works. Keys to success include using compost to cover the soil; replacing a harvested plant with a seedling immediately, which prevents weeds from getting established; building soil carbon quickly; and making a home for beneficial insects, which means growing hedgerows—lots of hedgerows. Paul Kaiser knew from his experience in Africa that if you wanted to have beneficial bugs on your farm, you needed to give them a home. In fact, the Kaisers have planted over three thousand native perennial bushes on the farm, because hedgerows are the perfect habitat for bees, wasps, and aphid-eating ladybugs.

Another key is the CSA. Their customers are loyal and generous. My visit to Singing Frogs coincided with delivery day. Boxes were being filled with oakleaf lettuce, red kale, carrots, rainbow chard, cabbage, broccoli, fennel, and Genovese basil, among other vegetables—all harvested one day before the official start of spring! Kaiser said 99 percent of the food in their CSA comes from the farm, excepting only potatoes, rice, and mushrooms, which they purchase from neighbors. Normally, other CSAs incorporate food from multiple farms, including some that are many miles distant.

It all adds up to multiple advantages for this innovative year-round model:

• Satisfied CSA members renew their memberships throughout the year, providing a continuous flow of cash to the farm;

- There is enough income to hire full-time employees and provide them with good benefits;
- There is a steady rhythm of work on the farm, based on 52 weeks of sowing and harvesting, instead of the traditional two big pulses in spring and fall;
- The year-round presence of beneficial insects means year-round pest control;
- And there is plenty of organic material for composting.

As if on cue, near the end of my interview an excited employee brought into the barn the longest earthworm anyone had seen on the farm. It was 10 inches at least. Everyone immediately took photos.

Now, you may be thinking "Come on, it's northern California. It'll never work where I live!" So I asked Kaiser for his take on that mind-set.

"The parameters of soil biology are the same the world over," he said. "The specific solutions to creating healthy soil are always contextual, but I've worked on the edge of the Sahara Desert and in the highlands of Costa Rica and it is possible everywhere."

Another question: Can the Singing Frogs model scale-up? In other words, can it feed a lot more people? It's a question that Kaiser gets frequently.

"I used to think that a one-hundred-acre farm like this would be great," he replied. "It's certainly possible, but a one-hundred-acre farm would probably have to be far from a city, so now I think many five-acre farms much closer to town would be the best way to scale-up."

And scaling up will be necessary, if we're going to intensify food production in order to feed an expanding global population sustainably. Fortunately, we have another innovative tool in the toolbox to accomplish this important task: never stop farming.

No wonder the frogs are singing!

TO LEARN MORE

Visit the Singing Frogs Farm website:
www.singingfrogsfarm.com

"The Drought Fighter," an article
about Paul Kaiser, is available online at
www.craftsmanship.net/drought-fighters

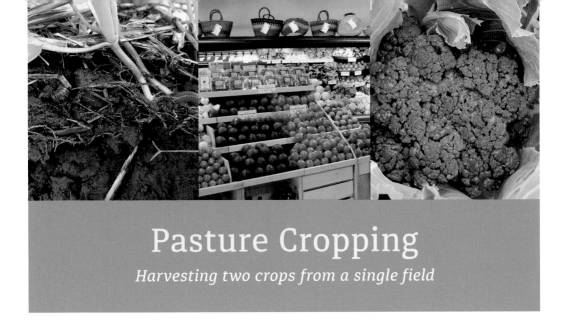

Pasture Cropping
Harvesting two crops from a single field

S ince the late 1990s, Colin Seis has been harmoniously planting
an annual cereal crop into perennial pasture on his sheep farm in
western New South Wales, Australia.

He calls this technique "pasture cropping." Seis gains two crops this
way from one parcel of land—a cereal crop for food or forage, and wool
or lamb meat from his pasture. It's all carefully integrated and managed
under his stewardship. And if he wanted, he could collect a harvest of
native grass seed as well, which was an important food source for the
Aborigines of the area.

In other words, pasture cropping accomplishes an important and
elusive goal: it intensifies food production for a growing population of
humans on the planet in a manner that is sustainable and regenerative.
This makes pasture cropping an innovative practice that sustainably
intensifies food production,
which means it has big potential
for feeding the world. Additionally, early research suggests
that it can also sequester loads
of atmospheric CO_2 in the soil.

However, the main reason
why today pasture cropping
is practiced by more than two
thousand farms across Australia is more mundane: it works.

The key to pasture cropping
is the harmonious relationship
between two different types of
plants: cool-season (C3) plants

Colin Seis raises two crops, oats and wool, from
this pasture-cropped field on his farm in New
South Wales. *Photo by Courtney White*

and warm-season (C4) plants (the "C" stands for carbon). Cool-season plants, such as wheat, rice, oats, and barley, grow early in the season and then become less active or go dormant as temperatures rise and light intensity increases. In contrast, C4 plants, such as corn, sorghum, sugarcane, and millet, remain dormant until temperatures become warm enough to "switch on" and begin growing.

Pasture cropping utilizes niches occupied by C3 and C4 plants. In winter, when C4 plants are dormant, the farmer uses a special planting technique involving a no-till seed drill to sow an annual C3 crop into a pasture of perennial C4 plants. With the onset of spring, the C3 seed germinates and begins to grow. With proper management and the right amount of rain, the C3 crop can be harvested before the C4 plants begin the vigorous part of their growth cycle. The removal of the C3 crop will then stimulate C4 plant growth due to reduced competition. Also, because shallow- and deep-rooted plants access water resources in the soil differently, overall productivity can increase.

A key is what's happening in the soil. Cool-season cereal crops provide sugars (food) to beneficial soil microbes, such as fungi, nematodes, and protozoa, during the time when the C4 plants are dormant. This improves soil fertility, because the more that soil microbes are stimulated to do their job, the better. This also speeds up nutrient cycling, promotes an improved water cycle, increases nitrogen content, and adds organic matter to the soil, which can build humus. Additionally, the no-till drill lightly aerates the soil, allowing oxygen and water to infiltrate.

Another key is using grazing animals to prepare the C4 field before drilling. By design, sheep or cattle graze the perennial pasture, cropping off the plants close to the soil surface, so that the C4 plants come up more slowly and give the C3 plants a chance to become established. By grazing heavily in the pasture with a large mob of sheep in a time-controlled manner, Seis can keep the C4 plants from growing too tall too early, and thus prevent them from shading the C3 plants. Animals can also control weeds, create litter on the soil surface, supply a pulse of organic nutrients for the crops, and remove dry plant residue from the pasture.

Harmony, easy.

Seis has some words of advice from his pasture-cropping

After this oat harvest is complete, farmer Colin Seis will turn sheep into the field to graze. *Photo courtesy of Winona Farm*

experience: use grazing to create as much litter on the ground surface as possible; sow the correct crop for your soil type; sow the crops up to two weeks earlier than usual because crops sown in a pasture-cropping setting are slower to develop; and avoid fertilizer use as much as possible. Seis cautions that crop yields are usually lower in the early years than with conventional agriculture. However, this is more than offset by the ability to produce two or more products from the same land, as well as the increase in fertility that is being built up in the soil.

When Seis took over the farm from his dad it was in trouble ecologically and financially. Pasture cropping turned everything around. Here are the benefits that it has brought him over time:

- It's profitable—Seis and his son run around four thousand Merino sheep and pasture crop around 200 hectares (500 acres) annually in oats, wheat, and cereal rye;
- The farm has steadily improved its sheep-carrying capacity, wool quality, and wool quantity;
- Once covered with nonnative plants, the farm has been restored to native grassland, with over 50 different species of grasses and forbs;
- The farm saves around $60,000 annually in decreased inputs (fertilizer, et cetera) in comparison to its former operation;
- Crop yields from pasture cropping remain about the same when compared to conventional cropping, with oat yields averaging 2.5 tons per hectare;
- Insect attacks and fungal diseases in crops or pasture are minimal;
- There has been a noticeable increase in bird and native animal numbers on the farm, as well as in species diversity;
- Soil microbial counts show that the soil has significantly higher counts of fungi and bacteria now than before;
- According to a soil analysis, all trace minerals and nutrients have increased by an average of 150 percent;
- Perhaps most impressively, soil carbon has increased by 203 percent over a 10-year span compared to an adjacent farm (owned by Seis's brother).

Christine Jones, an Australian soil scientist who has extensively studied the role of carbon in the soil, calculated that 171 tons of CO_2 per hectare has been sequestered to a depth of one-half meter on Seis's farm. This has contributed to a dramatic increase in the water-holding capacity of the soil, which according to Jones has also increased by 200 percent and is now more than 360,000 liters per hectare for every rainfall event.

In 2010, the University of Sydney conducted a research project on both Seis's farm and his brother's neighboring farm in order to evaluate the effects of pasture cropping versus conventional management on soil health and ecosystem function. The project compared paddocks of comparable size on each farm, discovering that there was greater ecosystem function on Seis's farm than on his brother's even though Seis's sheep stocking rate was double. They also determined that even though crop yields were the same on both farms, soil microbial counts showed that Seis's land had significantly higher amounts of fungi and bacteria.

In the study's conclusion, Dr. Peter Ampt and Sarah Doornbos wrote:

> *"These results illustrate that the rotational grazing and pasture cropping practiced on the innovator site can increase perennial vegetative ground cover and litter inputs, compared to the continuous grazing system and conventional cropping practiced on the comparison site. Increased perenniality and ground cover lead to improved landscape function in the pasture through increased stability, water infiltration and nutrient cycling which in turn can lead to improved soil physical and chemical properties, more growth of plants and micro-organisms and an ultimately more sustainable landscape."*

Although Seis's farm isn't certified organic, there's no reason pasture cropping can't be done organically, thus adding value to both the cereal and animal products. It can even be done with draft animals, if one chooses. Whatever method, the bottom line is the same: healthy soil. Or as Seis likes to say "The best way to improve your profits is to improve your soil."

That sounds like a harmonious combination to me!

TO LEARN MORE

"Communities in Landscapes Project: Benchmark Study of Innovators" by Peter Ampt and Sarah Doornbos. University of Sydney, 2011.

Colin Seis gave a lecture on pasture cropping at the 2012 Quivira Coalition conference: www.youtube.com/watch?v=h_4SrFlzclM

Sustainable by Tradition

Ancient farming practices with a bright future

There's nothing new about regenerative agriculture—except almost all of it.

It is easy to forget that once upon a time all agriculture was organic, grassfed, and regenerative. Seed saving, composting, fertilizing with manure, polycultures, low-till, and animal power—all of which we associate today with sustainable food production—was the norm in the old days, not the exception as it is now. Somehow, we managed to feed ourselves and do so in a manner that was self-reliant and self-perpetuating. You know what happened next: the plow, the tractor, fossil fuels, monocrops, nitrogen fertilizer, pesticides, herbicides, fungicides, feedlots, animal byproducts, *E. coli*, CAFOs, GMOs, erosion, despair—practices and conditions that most Americans today think of as "normal," when they think about agriculture at all.

Fortunately, deep-rooted agricultural traditions persist in many places around the planet and can help us meet modern food and water challenges. Two examples are landraces and the milpa.

Landrace is a botanical term for the locally adapted variety of a crop that has been grown by farmers in a particular community or region for a very long time. Landraces are resilient to adverse climatic conditions because they have been associated with their local soils (and farmers) for so long that they have seen it all: late frosts, hailstorms, drought, floods, pest attack, and competition with weeds. Over the years, these events have "thinned the herd" of the crop's genetic base, leaving only the hardiest members to propagate future generations. By saving the seeds of resilient parent plants, traditional farmers cultivate adaptable strains of crops that grow well under a wide range of difficult climatic conditions—which sounds very useful in the twenty-first century.

Miguel Santistevan teaches young people how to farm in the centuries-old traditions of northern New Mexico. *Photo courtesy of Taos Land Trust*

I learned about landraces from Miguel Santistevan, a dynamic teacher, farmer, activist, and researcher. Born and raised in northern New Mexico, Santistevan's list of credentials is impressive: a PhD candidate in biology at the University of New Mexico, permaculture specialist, public radio show producer, mayordomo for a local acequia (water ditch) system, heirloom seed saver, and farmer on his family's land. The tie that binds all of these activities together is his deep attachment to the ancient acequia-irrigated and native dryland agricultural systems of the Upper Rio Grande region.

One of his main passions is mentoring young people in traditional (sustainable) agriculture, reminding them through hands-on experience that what's new is actually quite old. Every year, he teaches groups of high-school-age youth how to grow food according to New Mexican traditions, including Hispanic agricultural practices that date back four hundred years. He guides the students through the process of picking and planting seeds and nurturing and harvesting crops, focusing on age-old staples such as corn, beans, squash, chiles, and melons.

The landrace varieties of these crops are the subject of Santistevan's doctoral research and the key, he believes, to agriculture's future in an era of hotter and drier conditions—conditions, by the way, that have already developed, as Santistevan and his young farmers know from their own hard experience.

"Our youth are our seeds," Santistevan said, "but what kind of conditions are we creating for them to germinate into the future?"

As an illustration of his concern, in 2011 the snowpack in the mountains above Taos was very light, causing the acequias that fed the farm fields to go dry by the third week in June. Since Santistevan and his students don't use water from underground wells, they had to rely on an old-fashioned technology to keep their crops alive: watching the clouds (and regular visits to the Weather Channel). No substantial rains arrived, however. Instead, it sprinkled rain now and then and a few drops made their way down the plant stems to the roots, but that was all. It was enough, as it turned out.

"Our *alberjon* [peas] were flowering and were able to set seed in the next several weeks without substantial water," Santistevan said. "The *maíz blanco* [white corn] looked shorter than usual with stressed leaves, but when it was all harvested, we had several ears of corn that were obviously unaffected by water shortage, confirming all my beliefs in the potential resilience of this ancestral staple. Some other crops, such as lentils and fava beans, shriveled up in the heat as if they were burned under a magnifying glass. Surprisingly, we were still able to bring in a few dozen seeds of each."

For all the challenges facing the Southwest and its food producers, Santistevan remains optimistic. That's because the region is home to cultures and crops that are well adapted to the extreme and uncertain climate patterns.

"We never use a prediction of drought to deter our plans of planting," he said. "Rather, we embrace the difficulties as an opportunity to discover the 'champions' in our crop populations while we hone our techniques to learn how to meet the challenges of drought and climate change."

The second example of traditional practice that has stood the test of time is the milpa, an unplowed plot of farmland planted with corn, beans, and squash, often referred to as the Three Sisters.

As indigenous tribes have known for centuries, these three crops complement each other ecologically and nutritionally. Maize (corn) requires high levels of nitrogen in the soil to grow properly. Bean plants have a special ability to draw nitrogen from the atmosphere into the soil, and the maize repays the debt by providing stalks for the bean plants to climb. Squash grows between the maize rows, acting as a mulch, which helps to keep weeds down. Often the borders of milpas are planted with chiles, as a pest control, interspersed with melons. In this way, space is saved, water conserved, and labor reduced. Milpas can be fertilized by compost or other organic material, such as ashes from kitchen fires and manure, which enrich the soil. On tiny amounts of land, the bounty can be plentiful. If milpas are planted with landrace varieties of crops, then the best of both worlds come together.

Milpas can involve groupings other than the Three Sisters, too. In the tropics of Central America, milpas involve trees and have been an important part of traditional subsistence agriculture in the region since the days of the Maya. There are four stages to a typical tropical milpa, spread out over twenty to thirty years. First, the forest is cleared, burned, and planted. For a few years, the Three Sisters grow in full sun.

The forest milpa is a traditional form of agriculture involving corn, beans, and fruit trees. This one is located in the state of Quintana Roo, Mexico. *Photo by Macduff Everton*

Below them are herbs, tubers, and other plants cultivated to reduce pests and enhance the soil.

Second, the milpa evolves into a forest garden. Quick-yielding fruit trees, such as plantain, banana, and papaya, are planted among the corn, beans, and squash, followed by avocado, mango, citrus, and guava trees. Third, the fruit trees mature, producing fruit and creating a new canopy, blocking the sun and inhibiting undergrowth. The Three Sisters go out of business. Hardwood trees, such as cedar and mahogany, are planted.

Fourth, the forest garden becomes a hardwood forest. The farmer lets the hardwoods grow tall and create a high canopy. Eventually the farmer or his family will harvest them, then clear and burn the forest and start a new milpa all over again.

There's a lot more to a milpa, including complex social and cultural relationships involving their cultivation, but essentially it's a sustainable way to maintain wildlife habitat while producing plants for food, spice, shelter, medicine, and profit.

Landraces and milpas remind us that as we consider solutions to pressing problems, we should look to time-tested traditions that have sustained populations of humans for many generations.

TO LEARN MORE

For more on Miguel Santistevan's work,
see: www.solfelizfarm.wordpress.com

For a global perspective on ecoagriculture,
see: www.peoplefoodandnature.org

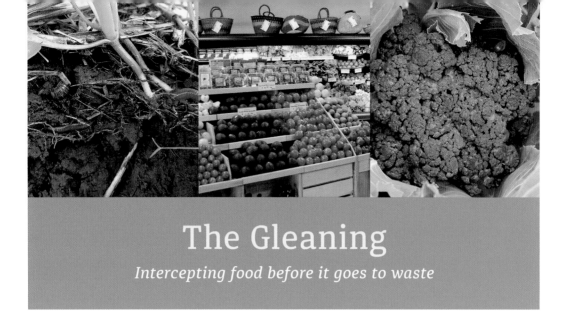

The Gleaning

Intercepting food before it goes to waste

A quick path to a sustainable and regenerative planet is also one of the simplest: we could stop being so wasteful.

Take food, for instance. According to various studies, between 20 to 40 percent of all food served in America goes uneaten and winds up in a landfill. That's approximately 83 *billion* pounds of food, or 270 pounds per person per year, worth at least $165 billion (in 2008 dollars). Not only is that outrageously wasteful in a world rife with hunger, when food rots it produces methane—a potent greenhouse gas—which means that Americans are producing *a lot* of methane from landfills just from our leftover food.

These numbers don't even include unharvested produce in the field or grove. By one estimate, 7 percent of the produce grown each year on American farms gets left behind. Sometimes that's because farmers grow more crops than they can sell as a hedge against disease or bad weather, sometimes there are labor shortages, but often produce is stranded because it can't meet industry standards for color, shape, or size. And when that happens, all the water, energy, and fertilizer (organic or otherwise) that went into making those crops went for naught, which adds to the wastefulness of our food system.

In the meantime, nearly 50 million Americans live in food-insecure households, including 15 million children, according to Feeding America, a nationwide nonprofit network of food banks. Food insecurity is defined as living without reliable access to a sufficient quantity of affordable, nutritious food at the household level, usually as the result of short-term financial distress. Then there's poverty—including rural poverty—which is a chronic condition and often means that members of a household lack an adequate supply of nutritious food on a year-round basis.

Fifty million Americans live in food-insecure households, including 15 million children. Gleaning fruit from backyard trees can help end hunger. *Photo courtesy of Portland Fruit Project*

And the very poor often eat very poorly.

One practice that can help is gleaning. It's the ancient practice of allowing the poor and disenfranchised into fields, orchards, and vineyards after the harvest to collect any remaining crops. Not only was it an efficient way to feed people, it has a religious mandate as well. The Bible's instructions to farmers are thus: "When you reap your harvest and forget a sheaf in the field, you shall not go back to get it; it shall be left for the alien, the orphan, and the widow, so that the LORD your God may bless you in all your undertakings" (Deuteronomy 24:19).

Go forth and glean. And give the food gleaned to the needy—which is an important component of the charitable tradition of Christianity and other religions.

These altruistic motivations are part of the reason why there's been a nationwide resurgence in gleaning, which is defined today as the collection of nutritious, locally grown food that otherwise would go to waste, including its redistribution to the poor and hungry. This good work is largely done by volunteers under the direction of a community food bank and coordinated with regional food security and social justice networks, all with the goal of reducing waste while feeding hungry people. It is a simple blessing for the planet and its people.

Gleaning connects the dots between surplus food and hungry people, especially in cities that have fruit trees or other edible plants. I first heard about gleaning as a teenager growing up in Phoenix, Arizona. While riding my bike, I often saw citrus trees groaning with unpicked fruit hanging over backyard walls, creating a messy obstacle course on sidewalks. When I learned that volunteers had begun harvesting citrus (and Phoenix had *a lot* of backyard trees) and taking the fruit to the local food bank for redistribution to homeless people, I cheered.

A good example of this type of work is the citrus-gleaning program of the Community Food Bank of Southern Arizona, which serves the greater Tucson area. If a landowner wishes to participate, he or she schedules an appointment and then paid staff and volunteers come to the home or office to pick oranges, lemons, and grapefruit from the trees. Four or five houses can be done in a day. Alternatively, land-owners can do the gleaning themselves and drop off the fruit at a local branch of the food bank, or they can bring it to one of the food bank's "Citrus Saturday" events.

Either way, since the gleaning is done usually in the winter, Tucson's abundance of citrus trees provides a well-timed source of food for the area's neediest citizens.

Up in Oregon, which has a decades-long tradition of gleaning (sparked by a tax credit for farmers created by the Oregon State legislature in 1977), the Portland Fruit Tree Project employs two sig-nificant variations on the Tucson model: volunteers get to keep half of what they pick, and 50 percent of the gleaning teams are reserved for able-bodied low-income residents. The remainder of the gleaning is donated to low-income people with physical challenges or to food banks and other charities. This Project has been steadily successful and new gleaning groups have recently formed in Salem, Corvallis, Eugene, and Roseburg, Oregon.

Over in Vermont, the focus is on farms. The Intervale Gleaning and Food Rescue Program, located in northern Vermont, mobilizes volunteers to gather surplus food from area farms during the growing season and redistribute it for free to needy families and social service agencies. The program is part of the Vermont Gleaning Collective, a nonprofit that promotes farm-gleaning programs across the state and provides educational outreach to the public about the benefits of sus-tainable agriculture.

The Collective is spearheaded by Salvation Farms, a nonprofit food hub located in Morrisville whose goal is to build increased resilience in Vermont's food system through what they call "agricultural surplus management"—gleaning, in other words. Because there is more food left behind on farms than volunteers can pick each year, an intriguing effort is underway to create gleaning crews of prisoners from a state correctional facility, which would also provide training to inmates in sustainable agriculture.

Nationally, there's the Gleaning Network, a project of the Society of St. Andrew charity, headquartered in Virginia. The Society bills itself

Volunteers with the Portland Fruit Tree Project get to keep half of what they pick and 50 percent of the gleaning teams are reserved for able-bodied low-income residents. *Photo courtesy of Portland Fruit Tree Project*

as America's "premier food rescue and distribution ministry" and has the numbers to back up its claims: since its founding in 1983, the project has gleaned 700 million pounds of produce, including trailer-loads of potatoes rejected by growers for their slight imperfections. Over 25,000 volunteers do the work each year, with an average of 15 gleaning events taking place every day somewhere in the nation. The Society also operates a training program for youth called Harvest of Hope.

All of this is a great example of how the "better angels of our nature," to quote Abraham Lincoln, can be put to work by a small number of people for the benefit of the wider community. It's not simply altruism at work either. This is important to keep in mind as natural resources, and the life that depends on them, become increasingly stressed this century. Here's biblical advice to heed on that score: "Give, and it will be given to you. Good measure, pressed down, shaken together, running over, will be put into your lap. For with the measure you use it will be measured back to you" (Luke 6:38).

What goes around, comes around!

TO LEARN MORE

"Wasted: How America Is Losing Up to
40 Percent of Its Food from Farm to Fork to Landfill"
by Dana Gunders. Natural Resources Defense Council.
NRDC Issue Paper, August 2012.

For more about the Portland Fruit Tree Project,
see: www.portlandfruit.org

For more about the Gleaning Network,
see: www.endhunger.org/gleaning-network/

Let's Glean! United We Serve toolkit is available at
www.usda.gov/documents/usda_gleaning_toolkit.pdf

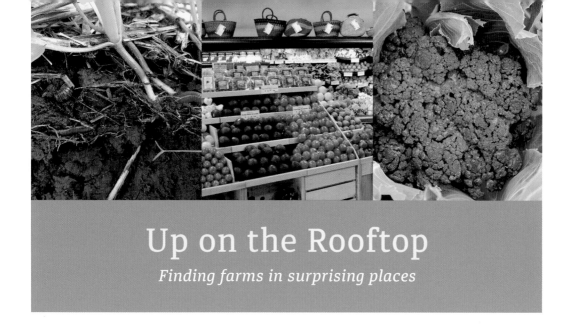

Up on the Rooftop
Finding farms in surprising places

I had never climbed three flights of stairs before to visit a farm.

But that's what I did after emerging from a subway station in Brooklyn, New York, and walking along Eagle Street to a warehouse owned by a television company called Broadway Stages. Up on the roof, I saw hundreds of vegetables set in neat rows of dark, rich soil. Walking to the edge, I saw the East River and beyond it a sweeping view of midtown Manhattan and the Empire State Building.

Wow.

I had come to Eagle Street Farm to see the nation's first commercial rooftop farm in action and to meet Annie Novak, the farm's charismatic cofounder, a Chicago girl who grew up reading *Vogue* magazine with dreams of "being fabulous" in New York City, and then became a truly fabulous rooftop farmer.

After graduating college, Novak landed a seasonal job at the New York Botanical Garden teaching children how to grow food. In the years that followed,

Annie Novak is cofounder and director of Eagle Street Rooftop Farm in Brooklyn, New York. *Photo by Avery Anderson-Sponholtz*

she balanced her city job with farming upstate, starting a nonprofit organization, and dabbling in the restaurant business. Eagle Street happened when the owners of Broadway Stages decided to install a green roof. Originally the plan had been to create an ornamental plant nursery, but Annie and cofounder Ben Flanner convinced the owners to give veggies a chance instead. They added compost to the soil mix, planted crops they knew were tolerant to heat and water stress, studied weather forecasts, and crossed their fingers. It worked.

Today the farm grows a wide range of crops, specializing in heat-loving and dry-tolerant chiles. The farm also keeps bees, rabbits, and hens. It sells its produce on-site and to local restaurants.

It hasn't all been a bed of roses, however. At times, windstorms and unseasonable heat bedevil both the veggies and their handlers. Space is a limitation—Novak can't expand the farm even though she would very much like to. In the beginning, fertilizer was another challenge because it had to be hauled up the stairs. The economics of rooftop farming are difficult too. The for-profit farm relies on value-added products like its hot sauce, called Awesome Sauce, to raise the $1.50 to $3 per square foot needed to farm unprocessed crops. At 6,000 square feet, farming at that scale makes just enough income to support a few part-timers, management included. But for impact far beyond its size, the economics are not as important as Eagle Street's educational purpose. And there Novak has found an eager audience.

"For folks who have nine-to-five jobs," she said, "it's nice to be able to come down on the weekends and get up to their elbows in dirt. One Sunday, all we did was carry up hundreds of garbage cans of soil to the roof. People were having a great time, spreading it like brownie mix. It's the hardest work you could think of, but people loved it."

Given the farm's small size, the most frequent question Novak gets is, "Can New York City feed itself?" Her response is unexpected: "Does New York want to?" She thinks not. "The quality of our air and water is protected by upstate organic growers," she said. It's important to her that those farmers, and the watershed in which they work, be supported by New York City residents.

Eagle Street has also inspired others to give rooftop farming a try.

In 2010, a group of young farmers formed a for-profit organization called the Brooklyn Grange and opened what has become the world's largest rooftop farm, located on two separate roofs in Brooklyn and Queens, totaling 2.5 acres. They grow more than 40 thousand pounds of organic produce a year, with tomatoes being the biggest crop. Their goal is to create a fiscally sustainable model for urban agriculture while

Brooklyn Grange farm covers 2.5 acres of rooftops and produces over 40 thousand pounds of organic food each year. *Photo by Cyrus Dowlatshahi*

producing healthy food from what they call the "unused spaces of New York City."

"We believe that this city can be more sustainable," they wrote on their website, "that our air can be cooler and waterways can be cleaner. We believe that the 14% of our landfills comprised of food scraps should be converted into organic energy for our plants, and plants around the city, via active compost programs."

The work of Brooklyn Grange has quickly expanded to include egg-laying hens, a bee-breeding program, and a farm-training program for interns. They host thousands of New York City youth each season for tours and workshops, launched the New York City Honey Festival in 2011, and provide a unique setting for corporate retreats, dinner parties, and wedding ceremonies. They also tackle environmental challenges peculiar to metropolitan areas. With a grant from the city, Brooklyn Grange sited its second farm on the 65,000-square-foot roof of a building in the historic Brooklyn Navy Yard, which allows them to manage more than one million gallons of stormwater, reducing the amount that overflows into Brooklyn's open waterways.

All of this good news raises a question: How many rooftop farms could New York City accommodate? No one knows the answer. It's an engineering question, really: How many buildings can support hundreds of thousands of pounds of dirt on their roof? Many, I bet. In the meantime, rooftop farming continues to spread.

In 2013, it arrived in Boston with the launch of Higher Ground Farm, which occupies 55,000 square feet on top of the Boston Design Center, making it the world's second-largest rooftop operation. The brainchild of two young farmers, Courtney Hennessey and John Stoddard, the mission of Higher Ground is similar to what Annie Novak and the folks at Brooklyn Grange pioneered: make a dent in the urban heat-island effect with a green roof; help with stormwater management; reduce carbon in the air; increase access to fresh, healthy food; create habitat for biodiversity; and provide educational opportunities, as well as many other cobenefits.

It's all part of an exciting movement to grow food within city limits. There are now gardens in city parks and schoolyards, farms in empty lots in Detroit, vertical farms in Wisconsin, goats in Chicago, and on and on. After my visit to Eagle Street I tried to find out how many urban farms exist in the US—without luck. I did learn, however, that there were 20 million victory gardens during World War II, providing more than 40 percent of the nation's vegetables. I also learned that 75 percent of the American population now lives in or near urban centers, which means that the potential for urban farming is large, as is the potential for carbon sequestration in urban soils.

Of course, there are many nonpractical benefits of rooftop farming as well.

"When I'm on a rooftop all I'm doing is listening to the sound inside a tiny seashell and trying to hear a larger ocean," Novak wrote in an essay for *The Atlantic*. "If you live in a city, take advantage of it. Soak up the street smarts and the rush of city living that also embraces outdoors and fresh tomatoes. You have to grow a small plot with a big picture in mind."

More than half of the people on the planet now live in cities, which means there is a great deal of opportunity for someone—three flights up!

TO LEARN MORE

For more about Eagle Street Rooftop Farm,
see: www.rooftopfarms.org

For more about the Brooklyn Grange rooftop farm,
see: www.brooklyngrangefarm.com

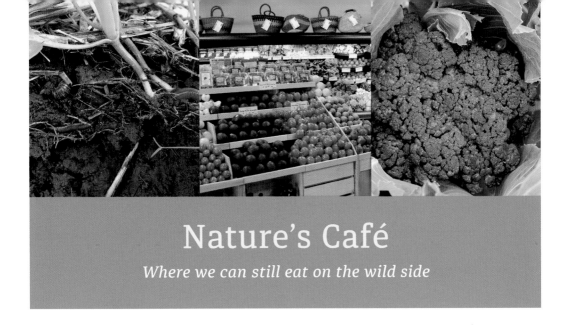

Nature's Café

Where we can still eat on the wild side

F ew questions have generated more books, articles, studies, lectures, fads, arguments, or confusion in recent years than this one: What should we eat if we want to be healthy?

We have been told to eat meat, to not eat meat, to eat only white meat, to eat mostly plants, to eat organic, to eat natural, to eat what our grandparents ate, to not eat genetically modified food, to skip carbs, to load up on carbs, eat less, eat more, to go vegan, go paleo, go South Beach, go Mediterranean, and on and on. It seems like a new set of instructions comes out every week, so it's no wonder that people feel bewildered.

Personally, I had settled on two simple answers: 1) If you are going to eat meat, eat only grassfed. 2) Eat more fruits and veggies, just like mom said, preferably from a local organic farm.

Recent research, however, indicates we should be asking a further question: *Which* fruits and vegetables? Specifically, which *varieties* should we be eating? New science says there are huge nutritional differences within types of fruits and vegetables. An apple is not an apple is not an apple, in other words. Some varieties will keep the doctor away, but some will make your doctor cringe with concern. That's because many popular apple varieties are badly deficient in nutrients and highly loaded with sugar. The nutrient content of the Jonathan Gold apple, as an example, is much lower than a less-widely available variety called Heritage.

For Jo Robinson, a pioneering journalist who was one of the first to broadcast the good news about the health benefits of grassfed beef, the answer to the question about what to eat is scientifically clear:

Eat on the wild side.

Not the kind of wild experienced by farmers two or three generations ago either, but the *really* wild—as in plants that were first cultivated four hundred generations ago. Her thesis, which she explains in her book *Eating on the Wild Side*, is this: the energetic campaign by humans over the centuries to make wild plants more productive, attractive, appetizing, and easier to harvest has significantly diminished the quantity and quality of their nutrients, many of which are essential to our health. These changes are so big, in fact, that the fruits and vegetables we eat today are essentially modern creations.

"Compared with wild fruits and vegetables," Robinson writes, "most of our man-made varieties are markedly lower in vitamins, minerals, and essential fatty acids. . . . Most native plants are also higher in protein and fiber and much lower in sugar than the ones we've devised."

There's another huge difference: wild plants are much higher in phytonutrients, which are bio-based compounds that protect plants from insects, disease, damaging ultraviolet light, and browsing animals. According to Robinson, more than eight thousand phytonutrients have been discovered by researchers so far, and each wild plant produces several hundred. Many of these are potent antioxidants, which fight free radicals in our bodies, responsible for damaging our eyesight, turning cells cancerous, and increasing our risk of obesity and diabetes. Phytonutrients have also been shown to reduce the risk of infection, lower blood pressure, speed up weight loss, protect the aging brain, lower "bad" cholesterol, and boost immunity.

"We will not experience optimum health until we recover a wealth of nutrients that we have squandered over ten thousand years of agriculture," Robinson writes, "not just the last one hundred or two hundred years."

This is a reason why this area of research is so hot today—and big business. The supplement market has exploded with phytonutrients, including pills, energy bars, juice drinks, and powders. However, Robinson says we don't need to give money to the pharmaceutical industry to get phytonutrients back into our bodies. Instead, we can shop "with a list," as she describes it, at our local grocery store and farmers market for fruits and vegetables that resemble their wild ancestors as closely as possible. Better yet, we can grow these varieties in a garden of our own.

Call it eating at Nature's Café.

The original menu at the café was dominated by plants that were tough, bitter, dry, astringent, seedy, and mostly sugarless. It's little wonder that as the agricultural revolution began to take off 10 thousand years ago, early farmers worked hard to cultivate plants that were sweeter, more tender, starchy, and oily. Cultivated dates, figs, and olives were early additions to the menu. In short order, we added a long list of cereal grains, including wheat in the Old World, corn in the New

World, rice in Asia, millet and sorghum in Africa. Over time, thousands of new café items were introduced to customers, many becoming highly popular, such as coffee, farm-raised meat, and anything containing sugar. With the Industrial Revolution and the rise of food science, the menu changed dramatically once more, as did our health. As we loaded up on sweets, starch, and feedlot beef, our well-being declined proportionally.

Grafitti cauliflower has twice as many antioxidants as other varieties, which is why it's a great example of "eating on the wild side." *Photo courtesy of Jo Robinson*

We didn't just lose phytonutrients in the process, Robinson says, our food has been de-flavored as well, ironically enough. That's because the food industry selects for ease of transport and storage, uniform appearance, and high productivity (including resistance to pesticides), all of which have had a negative impact on our food's flavor.

In her book, Robinson details how we can fight back by selecting fruits and vegetables that are high in phytonutrients and other good-for-our-health qualities (describing what experts call a low-glycemic diet). She offers a basic food rule: shop by color. Fruits and vegetables that are red, orange, purple, dark green, and yellow are among the richest in phytonutrients. But there are exceptions, and not all colors are equal (think apples), which is why you'll need to shop with a list. Here are a few quick examples:

Lettuce: go as dark green as possible; *corn*: blue, red, or deep yellow; *potatoes*: purple or French fingerlings; *tomatoes*: cherry, grape, and currant; *crucifers*: purple broccoli; red cabbage; orange, green, or purple cauliflower; and any type of kale; *beans*: black, brown, or red (canned beans preserve nutrients well); *lentils*: all varieties; *berries*: blueberries, blackberries, strawberries, cranberries, and raspberries; *grapes*: red, purple, and black (Concord grapes pass the test); *stone fruits*: choose the most ripe, shop for color, eat the skins, and go for the Bing.

I would only add that if you are an organic farmer, consider *planting* these crop varieties as well. Chefs and customers at Nature's Café need these ingredients in order to prepare their meals, but they can't eat them if no one plants them.

Liberty apple tree. A basic rule: shop by color—fruits and vegetables that are red, orange, purple, dark green, and yellow are among the richest in phytonutrients. *Photo courtesy of Jo Robinson*

Here's how Robinson answers the vexing question of what to eat to stay healthy:

"We can reduce our risk of disease by avoiding refined food and choosing high-phytonutrient, high-fiber fruits and vegetables that can restore a host of lost nutrients to our diet," she writes.

Put it together and we can have the best of all worlds.

"We can get additional health benefits by ramping up our physical activity so it comes closer to our long-ago ancestors," she concludes. "We can choose grassfed meat, which is similar to wild game meat. And we can combine this with the best of twenty-first-century medicine and can once again be healthy."

The answer is easy: eat at Nature's Café—every chance you get!

TO LEARN MORE

Eating on the Wild Side: The Missing Link to Optimum Health by Jo Robinson. Little, Brown, and Co., New York, 2013.

The Eating on the Wild Side fruit and vegetable shopping list is available on Robinson's website: www.eatwild.com

Cooperative Behavior
Farmer-friendly private enterprises serving the public good

"Food for People, Not for Profit."

This was the original slogan of La Montañita Food Cooperative, which was founded in Albuquerque, New Mexico, in 1976 with three hundred member families. It echoes the sentiments of many member-owned cooperative associations (co-ops) that started up in the late 1960s and early 1970s, according to Robin Seydel, who has worked at La Montañita since 1985. The co-op was very much a "hippie" establishment in the beginning, dedicated to gaining access to food that was "off-limits" at the time, including organics, whole grains, and macrobiotics. La Montañita also threw early jabs at the industrial food system by offering workshops on the links between pesticides and cancer, among other concerns. This counterculture spirit extended to its organizational structure as well; La Montañita deliberately set itself up as an alternative to the corporate model of soulless profit making.

Fast forward nearly forty years and what was once counterculture is now mainstream, from the type of food sold to the co-op model itself.

There are 30 thousand cooperatives in the United States, accounting for two million jobs and $500 billion in annual revenues. *Photo courtesy of La Montañita Co-op*

Today, La Montañita has over 16 thousand member households, employs nearly three hundred people, manages six stores in three cities, operates a regional food distribution hub, and has returned more than $4.5 million to its members in patronage dividends since 1989. It is an active member of the National Co+op Grocers, which encompasses over 140 food co-ops representing combined annual sales of more than $1.5 billion and over one million consumer-owners. And as we all know, healthy, nutritious, organic, and sustainably produced food that was once hard to find is now widely available across the nation and has become a part of everyday eating habits.

This good news begs a question: Could other kinds of regenerative activities considered economically off-limits today—such as building soil carbon or restoring damaged ecosystems or feeding large numbers of people sustainably—follow a similar trajectory?

It's not a pipe dream. Cooperatives are all around us, including worker-owned manufacturing co-ops, depositor-owned credit unions, and agricultural marketing co-ops. Overall, there are nearly 30 thousand cooperatives in the United States, accounting for two million jobs and $500 billion in annual revenues. IRS-recognized categories of cooperatives include: *consumer cooperatives*, which are owned by the people who buy their products or use their services—REI is the nation's largest example; *producer cooperatives*, set up so that farmers and others can sell their products under one label—Organic Valley is perhaps the largest in the nation; *purchasing cooperatives*, in which businesses work together in order to be competitive with national chains—such as the National Co+op Grocers; and *worker cooperatives*, which are owned and run by employees—a good example is the Mondragon Corporation in the Basque region of Spain, one of the biggest cooperatives in the world.

Not surprisingly, the rise of co-ops is closely linked to the labor movement. The first successful cooperative was organized in 1844 in Rochdale, England, when a group of weavers pushed back against the tide of industrialism sweeping the nation by opening a store to collectively sell their products. They called themselves the Rochdale Society of Equitable Pioneers and they authored a set of principles still in use today (recently updated) by the International Co-operative Alliance. They include open and voluntary membership, democratic control, economic participation by members, autonomy and independence, education and training, cooperation among cooperatives, and concern for community.

Counterculture indeed!

While the consumer cooperative category dominates in the United States, the cooperative movement as a whole is gaining momentum.

In addition to supplying its 16 thousand member households, La Montañita Co-op sells fresh, local, and organic food to the public. It also operates a regional food hub that serves underprivileged communities. *Photo courtesy of La Montañita Co-op*

Recent research suggests why: the broad and diverse benefits created by co-ops make them resilient in a crisis. Credit unions, for example, survived the Great Recession of 2008 relatively unscathed because they viewed rampant mortgage speculation as contrary to the interests of their members. Many cooperatives focus on the essentials necessary to a healthy society: food, water, electricity, insurance, and finance. Their primary mission is to provide public services, not to act as engines for wealth accumulation. This public-service orientation is why it is not such a big leap to extend the cooperative model to ecological restoration, renewable energy production, and carbon sequestration.

Although its hippie roots have faded, an important element of the cooperative model that remains firmly countercultural is its communal ownership structure. Like 501(c)(3) nonprofits, cooperatives are a legally sanctioned form of private ownership in service of the public good. While they are profit making, they are not profit maximizing. This sets cooperatives squarely against the corporate model of doing business, whose overriding goal is to turn a small pile of money into a larger pile of money, to paraphrase author and farmer Wendell Berry. In contrast, cooperatives see money as a means to an end: creating an economy that supports rather than diminishes the greater public good.

Robin Seydel describes the difference between the cooperative and corporate models this way: the size of financial dividends paid to members by cooperatives is based on patronage (how many goods and services they purchase), not on the percentage of their investment in the business. This is a big part of the democratic appeal of co-ops.

There are many other reasons to support the cooperative model. For example, La Montañita pays a living wage—and did so before living wages became popular—and it provides an excellent benefits package. Its food hub, the Co-op Distribution Center, serves several hundred local producers in a three-hundred-mile radius around Albuquerque. It is farmer- and rancher-friendly, sending them the important message that they can count on the co-op to be there—explaining the unofficial motto of the cooperative movement, which Seydel describes as: "We were local before local was cool."

For organizations like La Montañita, another motivating philosophy is the belief that cooperative behavior is the key to healthy communities and thus a brighter future for all. This shouldn't be news—humans have profitably engaged in cooperative behavior for millennia.

According to business expert and author Marjorie Kelly, the rising cooperative economy is helping to reawaken an ancient wisdom about living together in community. "They represent a need that arises from an unexpected place," she writes, "not from government action, or protests in the streets, but from within the structure of our economy itself. Not from the leadership of a charismatic individual, but from the longing in many hearts, the genius of many minds, the effort of many hands to build what we know, instinctively, we need."

Kelly says it's no accident that this redesign of our economy is beginning at the local level rather than in Washington, DC. That's because this redesign is "rooted in relationships: to the living earth and to one another. The generative economy finds fertile soil for its growth within the human heart." When economic relations are designed in a generative way, she argues, they're no longer about command-and-control behavior. "Economic activity is no longer about squeezing every penny from something we imagine that we own," she writes. "It's about being interwoven with the world around us. It's about a shift from dominion to community."

Cooperatives are an important place to bring innovative solutions together and make them work economically. That's way cool!

TO LEARN MORE

Visit La Montañita's website: www.lamontanita.coop

"The Economy: Under New Ownership: How Cooperatives Are Leading the Way to Empowered Workers and Healthy Communities" by Marjorie Kelly. *Yes!* magazine, February, 2013 see: www.yesmagazine.org/issues

Redefining Local

Linking farmers and consumers via the virtual marketplace

W hat does *local* mean when you live on a remote farm or ranch? It's an important question because going local has significant benefits: it gives us access to fresh, healthy food, reduces our carbon footprint, and lessens our dependence on fossil fuels. It keeps money circulating in the local economy, where its multiplier effect can be large. It bridges the urban-rural divide and helps to build a sense of community. And by supporting family-scale farmers, ranchers, and other businesses, it also pokes globalization in the eye.

Good stuff, but when we talk about local, we almost always mean local from the perspective of a city resident—those products grown or made closest to a customer. Farmers markets are a good example. Local in their case means a radius around a point (the market) located in a city or suburb. This means that participation is limited to those farmers and ranchers who can afford the time and money to drive into town every weekend. However, if you live on a remote farm or ranch, especially out West where the distances to potential markets can be staggering, local looks very different. Without a Santa Fe or Denver or Portland nearby, how can an organic farmer or grassfed beef rancher participate in the burgeoning local food movement and reap its benefits?

Fortunately, the Oklahoma Food Cooperative has come up with an ingenious solution: redefine local to include the entire state. They do this in two ways, with significant help from the Internet. First, it is a producer and consumer cooperative—rural producers and urban consumers gathered under one umbrella. Second, the buying and selling between the two groups happens in a virtual marketplace, which is where the Internet comes in. Here's how it works.

You pay a one-time fee of $51.75 to become a member of the cooperative. On the first day of every month, members can go on the

By linking customers and producers through the Internet, the Oklahoma Food Cooperative enables farms, such as this one in northwest Oklahoma, to sell their goods statewide. *Photo by Courtney White*

cooperative's website and purchase any food or craft product listed there. On the second Thursday, this electronic ordering "window" closes. The orders are then sent to the participating farms and ranches to be filled. On the third Thursday of the month, designated drivers visit all the participating producers to pick up the orders. All drivers then converge at a warehouse in Oklahoma City, where the products are separated into piles and then rebundled according to the customers' orders. The drivers travel back home, dropping off the individual orders at one of fifty designated locations across the state, where customers pick them up. Presto! Local redefined.

It is an impressive list of products available each month to members. There are over four thousand items on the Cooperative's web site, all made in Oklahoma, and many organic, natural, or grassfed. A sampling of items includes bakery goods, beverages, candy, canned foods, condiments, dairy and eggs, entrees, fruits, gift boxes, grains, flours and pastas, herbs, jams and jellies, meats, natural sweeteners, nuts, poultry, prepared foods, side dishes, and vegetables. Also apparel, art, baby products, bath and beauty supplies, books, classes, fiber arts, fishing supplies, health items, jewelry, laundry care, garden supplies, live plants, and seeds.

One downside to the Cooperative's model, however, is less face-to-face interaction between producers and customers. In both the community-supported agriculture (CSA) and farmers market models, the meet-and-greet relationship between grower and eater is an important part of doing business. By contrast, when they connect through the Internet, growers and eaters don't get much face time. For remote farmers and ranchers, however, this downside is offset by a big upside: they get to participate in a local food economy.

This model offers some other advantages for farmers and customers, too, some of which are also good for the planet.

- Each farm and ranch controls its own inventory and sets its own prices, and each designs its own label and controls its advertising;
- Member customers can order what they want, when they want it, and what they can afford, which means they are not locked in to the weekly produce list of, say, a CSA;
- Customers can earn credits toward a purchase by volunteering;
- Quality is guaranteed—or your money is refunded;
- For member producers, participation means making only one trip a month into town (when they are a designated driver) instead of the weekly trips required by the farmers market model;
- All participating farmers and ranchers get roughly 90 cents of every dollar paid for their products.

This last point is significant. In the industrial agricultural model, producers typically receive 20 cents of every food dollar. The rest goes to middlemen, including packers, truckers, grocery stores, and other corporate interests. In the Oklahoma Food Cooperative model there are no middlemen, other than the cooperative itself. Producers come out ahead because they are now "price givers" instead of "price takers." This is something new under the sun, which is why I made a long drive

John Gosney speaking to a tour group about how he gave up conventional farming to become a grassfed meat producer with the cooperative. *Photo by Courtney White*

a few years ago to Fairview, located in northwestern Oklahoma, to see for myself.

I wanted to see a farm in operation, so I joined a tour of Cattle Tracks, a certified organic wheat and grassfed beef farm owned by John and Kris Gosney. Originally, John Gosney was a conventional wheat farmer, soaking his fields with pesticides, harvesting the wheat with a ton of fossil fuel, and watching his spirit decline along with the land's health. He became depressed, he told the tour group, often finding himself sitting on a bale of hay wondering where his life was heading. Gosney said that he never gave organic agriculture a thought until a neighbor asked him to take over his farm because he was about to retire and didn't want to let his hard work developing an organic wheat operation come to naught.

After Gosney said yes, he was surprised to learn about the impressive profitability of his neighbor's farm. He decided to certify his own farm as organic as a consequence. He initially saw a drop in yield, but he also saw a drop in expenses when he stopped using conventional fertilizers and pesticides. Eventually, as the yield came up, so did his profits. But the main benefit of the switch, Gosney said, was noneconomic: he began to have fun again. Going organic cured him of his depression, he explained. He liked the challenge of organic as well as the hard work it required.

Now, the Gosneys grow cattle to eight hundred pounds on their fields and finish them on native grass (an all-wheat diet affects the taste of the meat). Gosney proudly pointed to an analysis by Oklahoma State University of the CLA (conjugated linoleic acid, a cancer-fighter) content of Cattle Tracks beef. According to the analysis it fell "in the highest range of CLA content reported in the literature for beef."

The Oklahoma Food Cooperative, he told us, was the key to it all. By offering products for sale via the Internet at a one-stop shop, the Cooperative extended the Gosneys' "local" customer base all the way to the state line.

Suddenly, remote doesn't seem so remote anymore!

TO LEARN MORE

To learn more about the Oklahoma Food Cooperative,
visit their website: www.oklahomafood.coop

For more information about
Cattle Tracks Farm, visit: www.johnsfarm.com

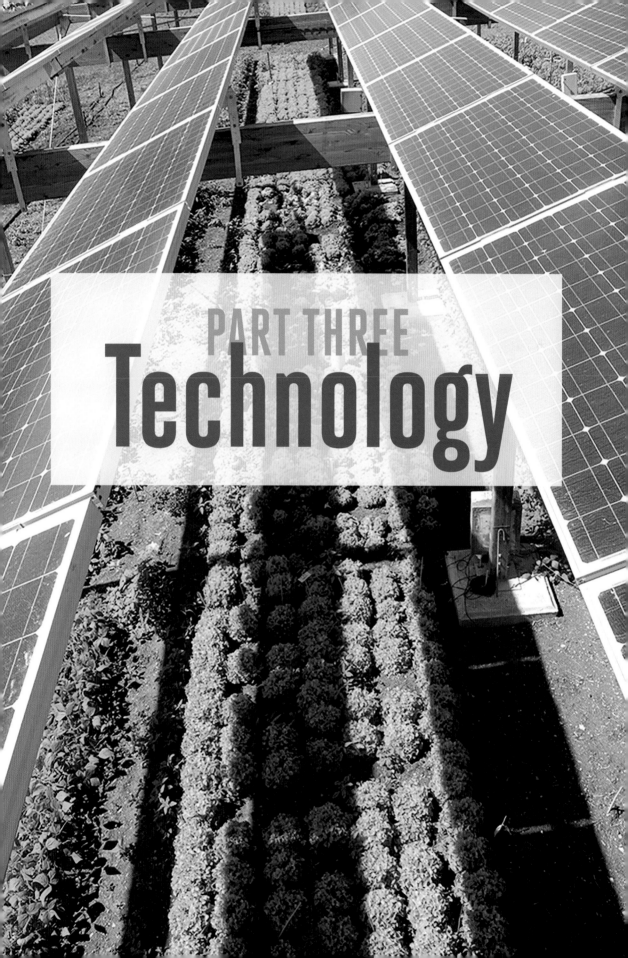

PART THREE

Technology

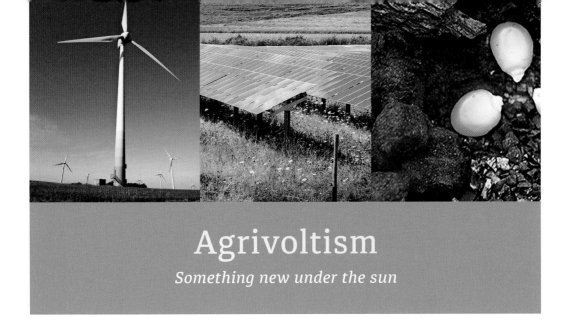

Agrivoltism

Something new under the sun

What is the best way to utilize sunlight—to grow food? or to produce fuel?

For millennia, the answer was easy: we used solar energy to grow plants that we could eat. Then, in the 1970s, the answer became more complicated as fields of photovoltaic panels (PVPs) began popping up around the planet, sometimes on former farmland. This was part of a new push for renewable energy sources, and as the technology improved over the years so has the scale of solar power projects on land that could otherwise produce food.

In the 1990s, the food versus fuel debate took a controversial turn when farmers began growing food crops for fuels such as corn-based ethanol, with encouragement in the form of government subsidies. Today the production of biofuels, including massive palm oil plantations, has become big business, often at the expense of hungry people and tropical ecosystems. As a result, the land requirement of the biofuels industry, not to mention its deleterious impact on ecosystems and biodiversity, has become huge—and it keeps growing.

Making the situation even more complicated and controversial is a simple fact: according to scientists, the amount of land needed to replace fossil fuels with biofuels exceeds all the farmland available on the planet. In other words, increased competition between food and fuel for agriculturally productive land means that the stage is set for food shortages and rising conflict as the projected human population on Earth swells to nine billion by 2050.

These developments led French agricultural scientist Christian Dupraz to ponder a question: Could food and fuel production be successfully combined on one plot of land? Specifically, why not build solar panels above a farm field so that electricity and food can be produced

This pioneering agrivoltism research site, located near Montpellier, France, produces renewable energy and food from the same plot of land. *Photo courtesy of Christian Dupraz / INRA*

at the same time? In addition to resolving the conflict between land uses, Dupraz hypothesized that solar panels could provide an additional source of income to farmers while sheltering crops from the rising temperatures and destructive hail- and rainstorms associated with climate change.

"As we need both fuels and food," he wrote in a scientific paper published in 2010 in the journal *Renewable Energy*, "any optimization of land use should consider the two types of products simultaneously."

In the same paper Dupraz coined a new word to describe this idea: *agrivoltaic*.

However, no one had implemented such an idea or researched its possible benefits and limitations. So, in 2010 Dupraz and his colleagues at INRA, France's agricultural research institution, built the first-ever agrivoltaic farm, near Montpellier, France, to test their hypothesis. In a 2,000-square-meter test field they planted crops in four adjacent plots—two in full sun (as controls), one under a standard-density array of PVPs (as if the solar panels had been mounted on the ground), and one under a half-density array of PVPs. The panels were constructed at a height of four meters (13 feet) to allow workers and farm machinery access to the crops.

The main issue was the effect of shade created by the PVPs on plant productivity. The researchers assumed productivity would decline, though there was scant data in the scientific literature to consult. That's why they built two different shade combinations, full versus half density, so that they could compare the effects of each with the other and of both with the control plots in full sun.

"Basically, solar panels and crops will compete for radiation," Dupraz wrote, "and possibly for other resources such as water, as solar panels may reduce the available water quantity for crops due to increased runoff or shelter effects." By the same token, shade can improve the productivity of crops in a warming world. "Water availability limits many crop productions. . . . Shade will reduce transpiration needs and possibly increase water efficiency."

As the experiment progressed, it became clear that a compromise needed to be struck between maximizing the amount of electricity produced by the solar panels and maintaining the productive capacity of the farm. Here's how Dupraz described it in scientific terms: "The optimum shade level for photosynthetic productivity would be one at which the level of photosynthetic photon flux density is high enough to saturate CO_2 assimilation but low enough to induce shade acclimation and reduce photoinhibition."

It was the Goldilocks principle at work again: too much shade hurt the crops, too little hurt electricity generation. Everything had to be just right. Could this balance be achieved? Variables the researchers identified included:

- The proper angle or tilt of the PVPs;
- The proper spacing between solar panels;
- Making adjustments for localized conditions (such as latitude);
- Choosing between fixed panels or panels on trackers (cost is a factor);
- The proper height of the PVP array;
- Engineering issues involved with the construction of the structure that holds the PVPs in place (the structures must be durable).

At the end of three growing seasons Dupraz and his colleagues had their answer: yes, balance was possible! But not quite for the reason they expected.

Not surprisingly, the crops under the full-density PVP shading lost nearly 50 percent of their productivity compared to similar crops in the full-sun plots. However, the crops under the half-density shading were not only as productive as the control plots; in a few cases they were even *more* productive.

The reason for this surprising outcome, according to Dr. Helen Marrou, who studied lettuce in the plots, was the compensating ability of plants to adjust to lower light conditions. She reported that in order to harvest light more efficiently, lettuce plants adjusted to decreased levels of radiation by an increase in the total plant leaf area and an increase in total leaf area arrangement.

Being good scientists, Dupraz and company were careful to say that more research was needed, including addressing questions about rain redistribution under the panels, wind effects on the crops, soil temperature changes, and the effect of dust from farming on PVP efficiency. However, their early results were very hopeful.

"As a conclusion," Dr. Marrou wrote in a paper, "this study suggests that little adaptation in cropping practices should be required to

Crop productivity does not diminish as expected under photovoltaic panels like these because the plants compensate for the shade. *Photo courtesy of Christian Dupraz / INRA*

switch from an open cropping to an agrivoltaic cropping system and attention should be mostly focused on mitigating light reduction and on plant selection."

To this end, Dupraz wrote me recently to say the next step in their research is to evaluate the advantages of using mobile solar panels mounted on trackers. This would allow them to adjust the radiation levels for crops to meet their physiological needs. It will also allow the panels to be tilted to a vertical position during rainfall events, giving the water a chance to fall uniformly on the crops.

Whatever the final disposition, clearly it's no longer an either/or situation. Thanks to the work of Dupraz and his colleagues, we know that agrivoltaic systems can combine food production with energy production on one parcel of land, while at the same time increasing the resilience of agriculture to climate change.

Which is definitely good news!

TO LEARN MORE

"Combining Solar Photovoltaic Panels and Food Crops for Optimizing Land Use: Towards New Agrivoltaic Schemes" by Christian Dupraz and associates. *Renewable Energy,* 36 (2011), 2725–2732.

"Productivity and Radiation Use Efficiency of Lettuces Grown in the Partial Shade of Photovoltaic Panels" by Helen Marrou and associates. *European Journal of Agronomy,* 44 (2013), 54–66.

The MoGro

Bringing healthy food directly to the people

It looked like an oasis in the middle of a food desert.

That's what I thought when I saw the mobile grocery store parked near the plaza at Santo Domingo Pueblo, a Native American reservation north of Albuquerque, New Mexico. Pueblo residents probably felt that way as well when the store-on-wheels made its first visit in the spring of 2011. A food desert happens when persistent poverty, inequality, and isolation severely limit nutritional options for the residents of a community. It's not just a rural issue either—there are many urban food deserts across the nation.

Physically, the mobile grocery—MoGro for short—is a large, custom-built semitruck that expands outward in the middle when parked. Inside is a full-service mini grocery store, including refrigerated sections for vegetables and frozen food. Flour, rice, milk, pasta, olives, meat, cheese, salad, canned goods—it's all there. With a catch. All of the food is either organic, grassfed, local, lean, or low-sugar (or a combination thereof), which means it's healthy. There's nary a can of soda pop, box of donuts, or bag of greasy potato chips in the whole place, and intentionally so. The food is affordable, too, which is another reason why the MoGro feels like an oasis. It's also popular, as I witnessed. It visits Santo Domingo Pueblo twice a week, attracting 70 to 80 customers per day. And much of what those customers purchase is fresh produce.

Philosophically, the MoGro is an attempt to address systemic health problems in Native American communities by providing culturally appropriate, nutrition-rich, affordable food on a regular schedule and with a convenient location. It's the creation of Rick and Beth Schnieders, who have an extensive background in food management, in collaboration with Johns Hopkins University's Center for American Indian

95

The MoGro is a mobile grocery store that delivers nutritionally rich, affordable food to people who live in food deserts. *Photo courtesy of Rick and Beth Schnieders*

Health, based in Albuquerque. The other partners are La Montañita Co-op and the tribal government of Santo Domingo Pueblo. Their collective vision is to eliminate food deserts by bringing healthy food directly to the people.

This is important because the nearest full-service grocery for many native communities is an hour's drive away, which means residents often choose the easier—and cheaper—alternatives for meals: fast-food restaurants and gas station convenience stores. Unfortunately, these kinds of highly processed meals have resulted in a well-documented epidemic of obesity, diabetes, high blood pressure, and heart disease among a wide swath of poor and disadvantaged populations. In particular, Native Americans have a 35 percent obesity rate, one of the highest in the nation. Their reliance on processed foods can be traced back more than a century to the time when the federal government, as part of its "acculturation" program, encouraged native families to adopt a western diet of lard, sugar, and white flour—food that completely upended traditional diets.

The idea for the MoGro took root when the Schnieders, who have been longtime supporters of the Center for American Indian Health, visited a grocery store in a remote part of the Navajo reservation. They were appalled by the food choices they saw. "They were all bad," Beth Schnieders told me. "There were no veggies in the entire store, for example." Meanwhile, the Center had received reports from its project workers that Navajo mothers were grinding up candy bars to feed their infants.

This gave Rick Schnieders a middle-of-the-night idea: a beer truck. Stocked with food, of course, not beer. At the time Schnieders was CEO of Sysco, the largest food-service corporation in America, which meant he knew a thing or two about food and food delivery. He also served on the board of Share Our Strength, an industry-supported nonprofit devoted to ending childhood hunger, which connected him with the Center for American Indian Health. And what the Schnieders learned from Center researchers was this: what native peoples needed more than access to better information and education was access to *healthy food itself.*

Enter the beer truck idea, now christened as the MoGro. The Schnieders and the Center entered into a dialogue with Santo Domingo Pueblo to see if tribal residents might be interested in their idea. The tribal government sent a survey to five hundred households and 98 percent of the three hundred household members who responded said they would be receptive to a mobile grocery. This kicked off a dialogue and planning process that lasted two years.

"They knew they had a problem," Rick Schnieders said. "One leader told us they were already building a dialysis facility for diabetes patients in the pueblo. They were definitely interested in alternatives."

In the meantime, the Schnieders addressed a technical question: What exactly *was* a mobile grocery store? When they looked around for examples, they found none. The closest prototypes were an unre-frigerated mobile store in Oakland, California, and the US military's mobile commissary for troops, which featured many nonfood items. Even an inquiry to a well-known think tank produced a dead end. Apparently no one had ever tried this idea before.

Their learning curve, in other words, was steep.

Working with Santo Domingo Pueblo and La Montañita Co-op, the Schnieders came up with an inventory of healthy, nonprocessed food that met the needs of tribal members. Next, they custom designed a semitruck from scratch. Then they hired staff and began twice-a-week runs to the pueblo. Most of the groceries were set up and sold outside the truck, which turned out to be a mistake. Sun, rain, wind, and dust were hard on both the food and the shoppers. The answer was MoGro 2.0, an air-conditioned truck where customers shop inside.

Another challenge has been a pleasant one—the popularity of the MoGro. Not long after deliveries began at Santo Domingo, the Cochiti Pueblo

A satisfied MoGro customer! By providing healthy food choices, the MoGro helps native communities fight diabetes and other widespread health-related challenges. *Photo courtesy of Rick and Beth Schnieders*

contacted the Schnieders and asked to sign up. Others followed. Today, the MoGro makes regular visits to San Felipe, Jemez, and Laguna pueblos as well. Additionally, it stops for a half day at the nonnative community of Cochiti Lake.

For all its pioneering fits and starts, the MoGro appears to be a success. Not only is it in demand—the Schnieders have fielded inquiries from all over the world—it has tangible positive impacts on the communities it serves. MoGro customers enjoy affordable access to healthy foods and foods that are part of local food traditions. And because of the "MoGro Bucks" discount programs, customers can save up to $100 per week. A less tangible but very positive impact is the reduction in carbon footprint: MoGro customers don't have to drive to the grocery in town. In total, that adds up to thousands of miles of driving avoided every week, 52 weeks a year.

So, is a MoGro replicable in other food deserts, including urban ones? Absolutely, say the Schnieders. The key to making the model work is:

- Community support (local leaders, local hires, cultural sensitivity);
- Regularly scheduled hours;
- Low prices;
- A central warehouse (La Montañita Co-op, in this case).

The MoGro faced various challenges too. Start-up costs were steep, although not as steep as the cost of building a grocery store. The team needed patience, because the MoGro took a while to break even financially. (The MoGro is run as a nonprofit, and fund-raising is a perennial challenge.) Meeting rising demand is a challenge, too, as is resisting pressure to include soda, candy, donuts, and other junk foods on the store's shelves.

Like any pioneering project, there's still a learning curve, but the Schnieders say the MoGro has turned a corner and should do fine on its own into the future.

Hopefully, this oasis is no mirage!

TO LEARN MORE

Visit the MoGro Mobile Grocery website: www.mogro.net

For more information about the Johns Hopkins Center for American Indian Health, see: www.jhsph.edu/caih

Farm Hack

The Internet meets regenerative agriculture

Welcome to the virtual coffee shop for agrarians!

Pull up a laptop and join the conversation. Do you have a farming issue on your mind, or maybe a tool design that you'd like to share, a crop problem that needs to be solved, a beginner's question that needs to be answered, or an intriguing idea about carbon sequestration that needs to be floated? If you do, Farm Hack is the place to go.

It's not the Bellyache Café, however. Leave all complaints, rants, and political opinions at the door.

This might be unusual for a web-based conversation site, to say the least, but a lot about Farm Hack is unusual, as I found out when I attended a Farm Hack meet-up in Hotchkiss, on Colorado's western slope. A small group of farmers, ranchers, and conservationists got together for a day to tackle the difficult topic of building drought resilience on the small-scale farm against the backdrop of rising water scarcity in the West. Not only is the region stuck in a persistent dry spell; long-range forecasting models that factor in hotter and drier conditions under climate change suggest that sooner or later reduced water supplies will become the new normal. If ever a subject needed a coffee-shop brainstorm, this was it.

The nonprofit Farm Hack bills itself as an "Open Source Community for Resilient Agriculture." It was born a few years ago during a design workshop at the Massachusetts Institute of Technology that involved engineers and young farmers, and it quickly evolved into an online platform to document, share, and improve farm tools. A quick peek at the website, for example, reveals how-to information on the benefits of a small axial-flow combine harvester (way cooler than it sounds), picking the right organic carrot seeds, building a pedal-powered

An example of Farm Hack innovation: the "culticy-cle" is a pedal-powered tractor for cultivation and seeding, built from lawn tractor, ATV, and bicycle parts. *Photo courtesy of Dorn Cox*

root washer, measuring soil carbon, and using low-cost balloon-mounted cameras for imaging a farm.

If that sounds more toolshed than coffee shop, Farm Hack is also where young farmers —including the young at heart—can start a conversation with experienced agrarians, skirting the need to reinvent various wheels on the farm. In addition, the site serves as a platform to share the latest research and make connections with like-minded individuals and organizations.

And you don't have to burn a gallon of diesel to get to this meeting place!

Farm Hack was incubated by the National Young Farmers Coalition (NYFC), a nonprofit founded in upstate New York in 2010 by and for a new generation of farmers in the United States. The NYFC is composed of new farmers, established farmers, farm service providers, good-food advocates, conservationists, and conscious consumers. It has an admirable list of goals, including conducting educational activities that encourage sustainable agricultural practices, providing farmer-to-farmer training, and encouraging cooperation and friendship between all agrarians. Farm Hack is an example of a program that embodies all three of these goals.

Accomplishing these goals in our modern age means embracing the open-source culture of the Internet, which is where Farm Hack comes in. The site is managed on the wiki model, which means it can be freely edited by registered participants and a wide variety of content can be easily uploaded for all to see and share. All it takes to register is a user name and password. The site is dynamic, flexible, and ever evolving, much like the young farmers movement itself. For new farmers, Farm Hack can be a godsend because of the pressure to quickly "get it right" in our challenging times. Accumulating sustainable farming experience over twenty years, for example, might not be practical in a world of rapid economic and ecological change. Moreover, the tools available to farmers have changed dramatically in recent years, especially software.

As one participant at the Hotchkiss meet-up put it, "Building spreadsheets has become as important as picking the right crops or watching the weather."

According to Dorn Cox, a young farmer from New Hampshire and one of the website's cofounders, the word *hack* comes from the tech world, where it means repurposing with the goal of taking control of one's destiny. With Farm Hack, the goal of the nearly one thousand website registrants is

Farm Hack cofounder Dorn Cox, who farms in New Hampshire, is a leader in applying new technology, open-source software, and the Internet to organic farming. *Photo courtesy of Dorn Cox*

to repurpose agriculture toward a regenerative model with farmer-to-farmer innovation sharing and problem solving. It is also their goal to engage nonfarmers in the conversation, including designers, engineers, policy advocates, and anyone else interested in building a resilient food culture.

"It's a return to an earlier model, when agricultural information was widely shared," Cox told me, "rather than locked up in obscure journals or inaccessible scientific articles as it is today. Just as the local coffee shop or diner serves as the hub for exchanging experiences, a virtual coffee shop and field walk is needed to facilitate relevant experiences."

Cox also regularly checks an online forum called Public Laboratory, which develops and shares ultra-low-cost technology. It's where he found a $100 balloon-mounted camera that floats 25 feet above the ground. The infrared images taken by the camera have yielded important data about his farm that would otherwise have been hard to collect. Cox believes that online forums such as Farm Hack and Public Laboratory, along with traditional cooperatives and collaborative research projects with other farms, are as important to modern farming today as walking the fields each day.

"The complexity of my farming operation would be unmanageable without them," he said. "I'm certain that open-source knowledge sharing will revolutionize agriculture just as Wikipedia has revolutionized the encyclopedia."

While free-flowing dialogue and unobstructed access to knowledge, innovation, and data are keystones to the young-farmers movement today—as are the advanced technology and social media they regularly

Farm-fresh chiles grown near Hotchkiss, Colorado, on display as part of a Farm Hack meet-up tour.
Photo by Courtney White

employ—just as crucial are their off-line, face-to-face equivalents: meet-ups, hacks, or hack-a-thons (if longer than one day). These get-togethers initially involved detailed discussions about tools, including laptops and smartphones, but have expanded recently to include land management strategies, such as how to cope with drought. Whatever the topic, meet-ups have always featured a cross-section of skilled people eager to share what they know and learn in turn.

"We are focused on attracting into our community not only farmers but those with other relevant skill sets," Cox said, "including engineers, roboticists, architects, fabricators, and programmers. It's those who live to build and make things work that are the key allies to turn ideas into tools and then into finished products."

There have been a dozen meet-ups around the country to date, including events in Vermont, Detroit, Minnesota, and New York City, on topics as diverse as how to grow small grains, utilize draft horses, improve soil health, and start a farming operation. Our job in Hotchkiss was to ponder the future of sustainable agriculture in the face of hotter and drier conditions promised by climate change.

Farm Hack can help not only by providing a platform for sharing innovative solutions but by keeping our hopes up through contact and dialogue—whether in a virtual coffee shop or the real thing!

TO LEARN MORE

Visit the Farm Hack
website: www.farmhack.net

More information about the agrarian
movement can be found at the National Young
Farmers Coalition website: www.youngfarmers.org

Microsize It

Small is still beautiful

Often lost in the discussions about how to scale up regenerative solutions are the many advantages of scaling *down*.

It's been over four decades since British economist E. F. Schumacher coined the phrase "small is beautiful" as a pushback against the gigantism that was beginning to take over economic thinking and practice at the time. Alas, the giants prevailed. Today, going BIG rules nearly every aspect of our lives, especially in agriculture. We even think big when we make plans for renewable energy: extensive fields of photovoltaic panels, giant wind farms, huge hydroelectric dams, and the like. Our language has been supersized as well. We talk easily about gigawatts and terajoules of energy, petagrams of carbon (one trillion kilograms), and zettabytes of data (one billion trillion bytes). Whew!

Perhaps that's why small is beautiful is making a comeback.

Take microdairies. For decades, the dairy business was the epitome of "get big or get out." According to the US Department of Agriculture, the number of dairy operations in the nation declined from roughly 650,000 in 1970 to 75,000 in 2006 (an 88 percent decline) with most of the losses occurring among small- to medium-sized operations. Over the same period, the average dairy herd size increased from about 25 cows to 120, and milk production per cow doubled. Meanwhile, the number of industrialized megadairies is on the rise. One of the largest in the US has 15,000 animals, while a "superdairy" in China is milking 39,000 cows! These animals are permanently housed indoors and fed a diet that in no way resembles a pasture. As you can imagine, there is nothing beautiful about these dairy factories.

In 2006, in a deliberate pushback against this trend, longtime dairyman Steven Judge founded a new dairy in Royalton, Vermont, with only four milk cows. Simultaneously, he opened a business called

Microdairies typically milk fewer than 10 cows and are a growing part of local, sustainable food systems across the nation. *Photo by symbiot/ Shutterstock*

Bob-White Systems to serve other micro-operations. Driven by a passionate belief in local sustainable agriculture, Judge had decided to turn to technology for answers to long-standing challenges for small producers. According to the Bob-White website, Judge's goal is "to create appropriately-scaled dairy technology and equipment that will provide micro-dairy farmers with the opportunity to sell safe, farm fresh milk and dairy products directly to friends and neighbors."

This new line of equipment includes small-scale bulk tanks, portable milking equipment, dairy supplies, and a micropasteurizer that Judge invented which can process milk on-farm so that farmers don't have to ship their product to an industrial processor. (Alas, he discovered the hard way that inventing the pasteurizer was far simpler than getting regulators to approve it.) All of this technology has made dairy farming much more practical than ever.

According to the Bob-White site, "A single four cow micro-dairy can produce 20 gallons of milk (or more) per day. This is enough to supply up to 60 families with safe and delicious farm fresh milk. Further, micro-dairies are humane and low impact, and can be conveniently situated on just a few acres, without the pollution, noise, and odor of large dairy farms."

And in a sign that downscaling is proving popular, between 2007 and 2012 the number of dairies in Vermont milking 10 or fewer cows rose 30 percent to 220—roughly a quarter of the state's total!

Another example of going small is a form of renewable energy using water on a tiny scale, called microhydro.

An example of a portable sawmill. Micrologging is handy for thinning overgrown forests, has low labor costs, and can provide a convenient source of wood. *Photo by bright/Shutterstock*

Think of the massive turbines in a hydroelectric dam and then think of a turbine that you could grip with your hand—it's exactly the same practice and principle, only at the scale of a household or small business. Of course, falling water has been a source of energy for human enterprises for centuries, including the iconic water wheel–driven gristmill. However, microhydro is different because it employs small, sophisticated turbine technology instead.

Two conditions are necessary to make a microhydro system work: a steady supply of water, such as a perennial stream, and a sufficient drop in elevation to turn the turbine. Typically, water is collected at the stream edge and piped downhill to the turbine and then returned to the stream farther down via additional pipes. A surprising amount of power can be generated from flows as low as two gallons per minute and a drop as short as two feet and it can be delivered efficiently as far away as a mile.

It's a continuous source of energy too, unlike solar or wind—even in winter, as long as the source of water is steady. And it's relatively cheap—all the basic equipment costs less than $2,000! And because the water is returned to its source, using a microhydro system to generate energy doesn't diminish a natural resource.

There are disadvantages, however, to microhydro: pipelines become more expensive and complicated the farther away you get from the water source or try to handle larger volumes of water; flooding can be a problem; and depending on your energy needs, it might be more cost-effective to go with a solar system instead.

However, if you have a good site for a microhydro system, there are many technologies available that will help you to utilize this clean, renewable source of energy, especially if your requirements are small.

And speaking of renewable resources, wood is another frontier for going small: micrologging.

Although portable sawmills have been around for decades, it's only recently that the technology has improved to the point where it is now both practical and profitable to cut trees sustainably in a wide variety of landscapes. In the old days, a lumber mill had to be disassembled, moved, and reassembled. Today, a micro version can be towed or loaded into a pickup truck, driven to a site, set up quickly, and begin producing lumber. Often, these portable mills cost less than the pickup truck itself and are tough enough to be used in many kinds of forested landscapes. With more than 10 million private owners of forested land in the United States, this could be a boon indeed.

With the advance in technology, a variety of benefits from portable mills are now available: they are handy for thinning overgrown forests for ecological purposes; they have low labor costs; they provide a convenient source of wood; they can run on biodiesel, which means they have a light carbon footprint, especially if horses are used to skid the logs; they can be used to reduce the risk of wildfire; they can easily handle small diameter trees, which are not attractive to the timber industry; and they can contribute to a local economy.

These new technologies—microdairying, microhydro, and micrologging—can all complement a holistic vision of people living in harmony on and with the land. In fact, a combination of the three could be the foundation of homesteading for the twenty-first century.

It's good to know that small is still beautiful!

TO LEARN MORE

For more information on microdairies,
visit the nonprofit arm of Bob-White Systems:
www.americanmicrodairies.org

Microhydro: Clean Power from Water
by Scott Davis. New Society Publishers,
Gabriola Island, BC, 2003.

Bigfoot

Technology to reduce our energy footprints

When it comes to energy use, it's best to wear small shoes.

What do I mean? Imagine you're walking across a beach or a snowy field. Look over your shoulder, and what do you see? Footprints the size of your footwear, representing the size of *you*. But what if those footprints represented the amount of energy you used that day instead? Or the amount of water you drank, or the distance food traveled to get to your plate, or the quantity of polluting CO_2 released into the atmosphere as a result of your various activities? How big would those footprints be then?

If you live in an industrialized nation, you'd almost certainly see evidence of very large shoes. And if combined (energy + water + food), your footprints would be *huge*.

This concept of an energy footprint is important for a farm or ranch (or any regenerative enterprise), because oversized energy feet could easily negate the regenerative work of growing food and building soil carbon. If farmers burn a lot of petroleum, for example, to go organic or to increase the capacity of their operations, they could be invalidating any "carbon-friendly" claims they've been making. They might also be spending more money than they would like. That's why so many farmers, ranchers, engineers, inventors, tech-types, and others are working hard to shrink their footprints. It also makes good business sense.

Important first steps involve technologies and management practices that *increase efficiencies*, *reduce waste*, and *promote conservation*. Two more big steps, however, can carry us even farther: 1) converting to biodiesel and 2) going solar.

Biodiesel is created by mixing vegetable oil or biolipids derived from animal fats with alcohol (typically ethanol or methanol) and adding a catalyst, usually lye. During the ensuing chemical reaction, the

Canola plants are a feedstock for biodiesel. They can be grown on many farms and their end product can be used in almost all types of diesel engines. *Photo by Chamille White/Shutterstock*

hydrocarbons in the oil bond with the alcohol to form bio-diesel, leaving glycerin behind as the main by-product—to be made into soap. The process was patented in 1892 by Rudolf Diesel, a German engineer who wanted to ignite fuel in an internal combustion engine without using a spark. He discovered that ignition can happen when fuel is injected into a highly compressed air mixture, and one of the original fuel sources he developed was peanut oil. However, Diesel's biofuel was quickly replaced by more abundant, though less efficient, petroleum products.

Biodiesel feedstocks that can be grown as crops include canola, came-lina, safflower, flaxseed, rapeseed, soybean, jojoba, sunflowers, palms, poppies, pecans, avocados, oats, mustards, coconuts, castor beans, olives, and various nuts. The conversion of feedstock to biodiesel is uncomplicated and can be accomplished with off-the-shelf technology.

In addition to being produced on-farm, biodiesel has multiple benefits. It can be used in a conventional diesel engine very easily and it's cleaner burning than petroleum-based diesel. Biodiesel is a natural lubricant, which makes engine parts last longer, and it's 100 percent biodegradable. There's no threat to human health from biodiesel and it's safe to store, transport, and clean up if it's spilled. Also, critically, the production and use of biodiesel can result in a positive energy balance; for every unit of energy used to make a gallon, as many as three units of energy are gained. This is important if a farm or ranch wishes to achieve a *net carbon energy balance*, to create more energy (as output) than it consumes (as input).

In terms of power and fuel efficiency, biodiesel is equivalent to its petroleum-based kin, though it "clouds" at cold temperatures, which reduces its performance. In another downside, it reduces acreage for growing cash crops, raising food-versus-fuel questions. However, an answer to the acreage problem can be found at your local restaurant: french fry oil! Straight vegetable oil can be employed in a diesel engine without much trouble. I once heard a presenter at a conference say that if a VW bug can run on french fry oil, a tractor can too.

Approximately four billion gallons of waste vegetable oil are generated in the US each year, mostly from restaurants, which means its potential as a fuel source is large. Besides its relative abundance, other advantages include that 1) it's free; 2) it's carbon-neutral, unlike biodiesel, which requires extra energy to make; 3) using it is a way to recycle waste; and 4) it's ideal for driving your ancient Mercedes-Benz around the farm (as the conference presenter showed us).

There are technical issues, however, with french fry oil. One involves buying the correct injector pump for your tractor. Others involve filtration and temperatures, usually requiring a two-tank system for the tractor (one for conventional diesel). It's also important to have a reliable source of used vegetable oil, such as a local brewpub or restaurant, which can be good business connections for the farm or ranch. Collected in barrels at the source, the vegetable oil needs to be stored for a few weeks so that the food bits can settle to the bottom, though the process can be speeded up by warming the oil. This energy could be produced by burning on-farm biomass in specialized ovens, or . . .

You could go solar.

Thanks to various tax incentives and a sharp reduction in the price of photovoltaic panels in recent years, there has been a veritable explosion of solar energy in the nation's agriculture sector. It's a natural fit: farms use a lot of electricity, resulting in bills that run into the thousands of dollars each month, and they have the rooftops and open space on the ground needed for large solar arrays—including installing them above farm fields. As the technology has improved, the cost-benefit ratio of

Farms use a lot of electricity, which is one reason why there has been an explosion of solar energy in the nation's agriculture sector. *Photo by symbiot/ Shutterstock*

photovoltaic systems has dropped to the point where it makes good economic sense to convert to solar.

There's another reason: the footprint of an average solar panel has shrunk considerably.

The best way to determine the footprint of an object or practice is by conducting a life-cycle analysis (LCA), which is an inventory of the material and energy inputs and outputs at each stage of a product's life or the full duration of a management practice. Originally developed by television, stereo, and refrigerator manufacturers to determine the cradle-to-grave costs of their products, LCAs have become a useful tool for all sorts of enterprises, including sustainable agriculture. The LCA of a solar panel will include its raw materials and parts (quartzite, silicon, steel, and so forth), assembly, transportation, installation, length of use, and disposal, including possible recycling. In natural settings, such as farming, LCAs include air, water, energy, biodiversity, and social components. As you can imagine, calculating an LCA gets complicated quickly!

LCAs are also a useful way to assess—and reduce—the amount of greenhouse gas emissions that are being generated from a particular location. There are three basic parts to this analysis: the cumulative energy use of an operation; the size of the ecological footprint; and the amount of methane produced by the belches and farts of livestock. The good news is that there are many LCA models to choose from, many designed for organic and grassfed farm and ranch enterprises. The bad news is that an LCA can take a lot of time and energy away from the actual process of producing food and sequestering carbon in the soil.

Whatever path you choose, the goal is the same: wear smaller shoes.

TO LEARN MORE

Here is a how-to guide for making biodiesel:
www.journeytoforever.org/biodiesel_make.html

For a useful overview of on-farm
solar systems, see the USDA publication
"Solar Energy Use in U.S. Agriculture: Overview
and Policy Issues" by Irene Xiarchos and Brian Vick.
USDA Office of Energy Policy, 2011.

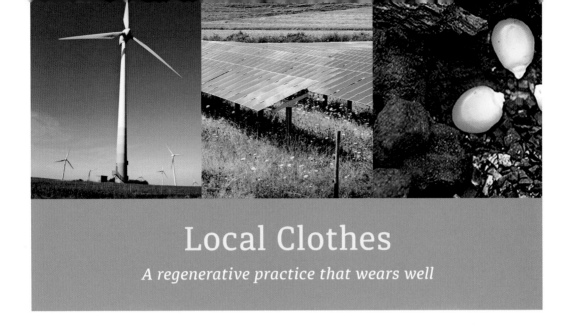

Local Clothes

A regenerative practice that wears well

O f all the human needs we strive to make sustainable, the one we consistently overlook is the one closest to our skin—our clothes.

It's an oversight we need to address, because almost everything we wear is drenched in fossil fuels, including the synthetic fibers that make up the majority of the raw material in clothes and the dyes that make them colorful. So, if behaving sustainably means procuring our food from a local foodshed and our water from a nearby watershed, why don't we try to procure what we wear from a local "fibershed"?

The quick answer is that we can't, because those locally made clothes don't exist. Not yet, anyway. However, Rebecca Burgess, executive director of the California-based Fibershed, and her partners are on it. If they have their way, someday we will be able to buy clothes made locally from natural fibers created by sustainable grazing and farming practices and spun in nearby mills powered by renewable energy, all part of a robust, low-carbon, climate-friendly regional economy. And that's just the beginning! Burgess envisions these fibersheds as

Rebecca Burgess, founder of Fibershed, wearing all-natural woolen clothes made within 250 miles of her home. *Photo courtesy of Fibershed*

111

the foundation for an international system of textile supply chains, designed to regenerate the natural systems on which they depend while creating a vibrant and lasting textile culture.

If that sounds utopian, well, consider the alternative: our current industrial system for producing clothes. Take water pollution, for example. According to the World Bank, textile manufacturing is the second largest source of freshwater pollution in the world (principally from dyeing) and accounts for 20 percent of all water contamination. Synthetic fibers, which make their way to the sea via sewer lines from industrial laundry operations, are a huge source of pollution in the world's oceans.

Those are just two of the environmental costs. Don't forget the low wages, terrible working conditions, and human rights abuses that are pervasive in the garment industry, including persistent slavery and child labor. The toll can be deadly. The collapse of a textile factory in Bangladesh in 2013 (despite warnings) killed 1,200 workers and injured more than 2,500 in the deadliest garment-manufacturing incident in history.

Burgess is quick to point out that the clothing industry is aware of these issues and that some larger corporations have begun to adopt eco-friendly practices, including the use of organic natural fibers. However, the goal of Fibershed is to go way beyond correcting deficiencies in the current system and create instead a radically new model, one inspired by time-honored traditions from around the planet.

The roots of the project go back to 2009, when Burgess decided to create and wear a prototype wardrobe made from fibers, dyes, and labor sourced within a 150-mile radius of San Francisco. To accomplish this goal, she pulled together a team of innovative agriculturalists and artisans to build the wardrobe by hand (because the manufacturing equipment had been lost decades ago). The team worked toward four specific objectives: produce no toxic dye waste; use no pesticides, herbicides, or genetically modified organisms; significantly reduce the carbon footprint of the wardrobe in comparison to conventionally produced clothes; and incubate a regional community of artisans and farmers that would collaborate and grow in number over time.

The prototype demonstration was a success on all levels, sparking widespread interest not only in the word *fibershed* (which Burgess coined) but in the concept behind it as well. To push the concept forward, in 2011 Burgess founded the Fibershed Marketplace to explore the possibility of creating a cooperative to help fiber farmers and artisans stay in business together. Then in 2012, she founded the nonprofit Fibershed in order to educate the public, including policymakers and

Diagram of an idealized woolen mill. *Image courtesy of Fibershed*

entrepreneurs, on the benefits of producing local clothes using regenerative practices.

Call it "thinking like a fibershed!"

Which raises a question: How is a fibershed defined exactly? According to Fibershed's website a fibershed is "a geographical region that provides the basic resources required for a human's first form of shelter (aka clothing)." However, don't get it confused with a watershed, warns Burgess, because a fibershed must necessarily cross multiple topographic boundaries to work ecologically and economically. Right now, that means stretching the definition of "local" way out—at least until sustainable fiber production takes off.

Another way to define a fibershed is to describe what's in one. The diagram on this page presents an idealized vision. It includes a solar-powered wool mill; a greywater dye garden; grazing sheep; industrial hemp, flax, and nettle cultivation; small-scale cotton-spinning equipment; a greenhouse; children visiting the field where their jeans are grown; a recycling mill; rooftop gardens for food, fiber, and dye plants; sewing pods; a knitting frame; and weaving studios.

It's a utopian vision that's very much grounded in reality.

For example, over three million pounds of wool are produced in California every year—more than anywhere else in the nation—of which 99 percent is shipped out of state, mostly to China. Much of this wool is wear-next-to-the-skin quality, which means that the raw material for the establishment of numerous fibersheds is already at hand. In fact, artisanal fiber operations have sprung to life in at least eighteen communities around the state since 2012, selling largely to upper-end markets. It's small, Burgess says, but it's a start.

A key component of Fibershed's work is its soil-to-soil concept, which aims to help ranchers and farmers build topsoil through a compost-application process that sequesters carbon dioxide on their land while

FIBER & DYE PROCESSING

clean, card, spin & dye fibers and weave or knit into fabric

provide fiber & dyes

SHEEP, COTTON, BAST FIBER & DYE PLANTS

DESIGNERS & MAKERS

SOIL to SOIL

design, cut & sew texiles/garments

provide nutrients

GARMENTS

RANGELAND, FARMLAND & CARBON SINK

apply to pasture and farmland

recycle the nutrients

COMPOST

This soil-to-soil life cycle of a garment shows its light carbon footprint and the use of sustainable practices. *Image courtesy of Fibershed*

reducing the product's carbon footprint. This is shown in the image on this page. "A typical wool garment produced overseas has a net carbon footprint of 33 kilograms in CO_2 equivalents," said Burgess. "The Fibershed approach reduces that and can, in fact, sequester nearly 38 kilograms in CO_2 equivalents per garment."

It's all bundled together in an idea called the California Wool Mill Project, which pulls together a broad array of regenerative solutions. The summary from the Project's feasibility study (available on the Fibershed website), which was conducted to assess the potential of producing cloth in a vertically integrated supply chain using 100 percent California-grown wool fiber, states that the goal of the Project is to create a technical road map for an ecologically sensitive closed-loop mill design utilizing renewable energy, full water recycling, and composting systems. Furthermore, the products from the mill were analyzed and shown to have a high potential for net-carbon benefit.

"The suggested model outlines the potential for a multi-stakeholder coop that would close the financial loop between profits and the producer community," wrote the authors, "furthering the positive economic impact for our ranching and farming communities."

In other words, we all live in a fibershed—we just don't know it yet!

TO LEARN MORE

To find out more about the Fibershed organization, visit their website: www.fibershed.com

A copy of the California Wool Mill Project Feasibility Study is available at: www.fibershed.com /wp-content/uploads/2014/01/Wool-Mill -Feasibility-Study-Feb2014.pdf

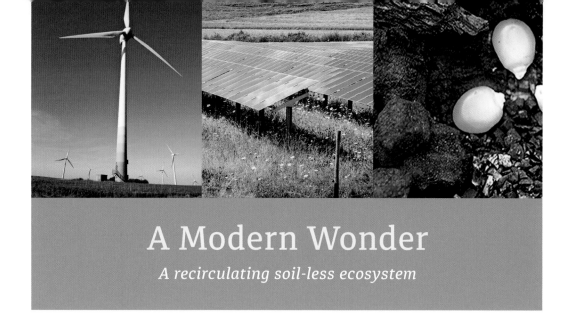

A Modern Wonder

A recirculating soil-less ecosystem

What highly productive, organic, regenerative food system can be built in your basement, bathroom, or abandoned pool and has the potential to feed millions of people?

Hint: it doesn't involve soil. Second hint: it does involve fish. Third hint: it might have been one of the Seven Wonders of the Ancient World.

The answer is: *aquaponics*. Not hydroponics. Not aquaculture. "Aquaponics"—an innovative blend of the two that overcomes each of their individual shortcomings, creating an easy-to-build, easy-to-maintain food system. Here's a definition provided by Sylvia Bernstein in her book, *Aquaponic Gardening*: "Aquaponics is the cultivation of fish and plants together in a constructed, recirculating ecosystem utilizing natural bacterial cycles to convert fish waste to plant nutrients. This is an environmentally friendly, natural food-growing method that harnesses the best attributes of aquaculture and hydroponics without the need to discard any water or filtrate or add chemical fertilizers."

Tomatoes meet tilapia. But let's back up for a second.

Hydroponics is a time-tested system for growing plants without soil. The plants are planted in trays of gravel or other nonsoil material and their roots are constantly bathed with nutrient-rich water. They can be raised outdoors under natural sunlight or indoors with grow lights. This system can be built in any configuration and, if well tended, can produce abundant yields in a short amount of time.

There are big downsides to hydroponics, however, including the high cost of the chemical nutrients. Toxic buildup of salt and chemicals in the water over time requires its periodic dumping and replacement, testing the water on a regular basis is a tedious chore, and going organic is difficult due to the reliance on chemical solutions to supply nutrients.

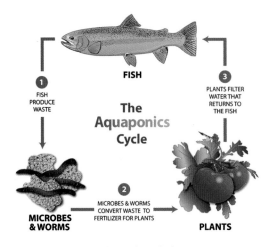

FISH

1
FISH
PRODUCE
WASTE

The Aquaponics Cycle

3
PLANTS FILTER
WATER THAT
RETURNS TO
THE FISH

2
MICROBES & WORMS
CONVERT WASTE TO
FERTILIZER FOR PLANTS

MICROBES & WORMS

PLANTS

The aquaponics cycle. Fish and plants are grown simultaneously in a healthy, efficient, recirculating ecosystem. *Image courtesy of The Aquaponic Source*

Aquaculture is a system for raising fish and comes in two basic types: offshore fish farms, either saltwater or freshwater; and on-land fish tanks. Both types are often maximized for production, which means the farms and tanks are often packed to the gills with fish, so to speak. Since raising fish is an efficient way to produce protein —it is less feed-intensive per pound than any other type of meat—aquaculture has become an important source of food for humans in recent years, especially as stocks of wild fish continue to be over-harvested around the world.

Fish farms, however, have serious waste-disposal and nutrient-management challenges, suffer from poor regulatory oversight in many nations, and are susceptible to killing viruses, which can spread quickly. Fish tanks also have major waste issues, which require frequent dumping of water. Additionally, the tanks are capital- and energy-intensive systems that require a lot of management and scientific research to keep the fish healthy. At industrial scales, these challenges are greatly amplified, which is why critics have dubbed them "fish feedlots," "fish factories," and "CAFOs for fish."

Which is where aquaponics comes in.

In an aquaponic system, water from a fish tank, which is full of nitrogen-rich fish waste, is filtered through a hydroponic plant arrangement and recirculated back to the fish tank. Round and round. The bacteria in the growing media love the fish waste, the plants love the natural fertilizer, and the fish love the clean water that comes back. This means that chemicals don't need to be added and water doesn't need to be dumped. And water use is minimal because it's a nearly closed system—the only water lost is what's taken up by the plants or evaporates. An aquaponic system can be built almost anywhere and in any configuration, including basements, bathrooms, and swimming pools.

According to Bernstein, an organic aquaponic garden can consume as little as one-tenth of the water used in a soil-based organic garden and can produce 50 thousand pounds of tilapia and 100 thousand pounds of vegetables per year in a single acre of space.

"Aquaponics can grow any type of vegetable and most types of fruit crops," she writes. "It can't scrub CO_2 from the air, but it can do almost everything else."

The system can easily be run on renewable energy, which means it won't generate green-house gases (unlike cattle, fish don't produce methane). And because it doesn't require soil, it doesn't need to compete with nature. "We don't have to clear a forest or a jungle to grow food,"

A basic backyard aquaponics system. Water from the fish tub is pumped up to the three garden tubs, where it percolates back down to the fish tub. *Photo courtesy of The Aquaponic Source*

she writes. "Instead, we can recycle old warehouses, derelict buildings, and other marginal places into farms."

Bernstein separates an aquaponic system into two parts: hardware and software. The hardware includes tanks, trays, pipes, pumps, and other equipment. These can be assembled into different formats, including vertical, raft, and nutrient film systems. Not only can the hardware be arranged to fit your space requirements, it can be inexpensive to purchase, especially if you recycle used materials (be mindful of their source, however).

The software includes fish, plants, worms, and bacteria. The system is gentle on all types of fish, though species that need warm or very cold water require more oversight. Tilapia are popular because they are hardy, grow fast, and can handle lower water quality, which makes them ideal for beginning aquaponic gardeners. Perch, bluegill, and certain types of bass are common, as are goldfish. Bernstein notes that you don't have to eat the fish—aquaponics is a great system for vegetarians too!

As for plants, almost anything will grow well, including broccoli, herbs, orchids, cactus, potatoes, carrots, shrubs, and flowers. "Sky's the limit" is how Bernstein puts it, though gardeners need to be mindful of pH and mineral balances.

The critical software, however, are the composting worms and the nitrifying bacteria, all of which thrive in the oxygenated water and rooting substrate. Their work in filtering the fish waste (transforming ammonia to nitrite and then to nitrogen) and transferring nutrients to plant roots is what brings the system to life and makes it regenerative. As for the substrate, Bernstein says the best kinds have "soil-like purposes," such as providing a stable foundation for the plants, a place for bacteria to grow, and a protective blanket against fluctuations in temperatures.

Put the hardware and software together and you have a system that can produce a great deal of fruit, vegetables, and fish protein rapidly and with relatively small space requirements. Since over half of the world's people now live in cities, Bernstein observes, aquaponics is an ideal way to feed many of them. Aquaponics is also ideal for schools, where it can be a great tool for teaching chemistry, ecology, and biology.

It all sounds great—but how many aquaponic gardens are there actually? Is this regenerative solution taking off?

Although the practice of aquaponics has been around for a long time, it's difficult to find hard numbers on how many systems are in operation today, possibly because of their inherent "backyard" nature. However, two indicators stand out: an Internet search reveals that there are many commercial suppliers of aquaponic parts and systems across the nation, suggesting there are a lot of customers out there; and second, aquaponics is part of the curriculum of many colleges and universities with sustainability departments, including Santa Fe Community College, near where I live in New Mexico. As for the science, a recent report on aquaponics by the UN's Food and Agriculture Organization examines these systems in great detail. It doesn't say how widespread aquaponics is globally, but I suspect there are more systems in operation than we realize. In any case, I'm certain it's a technology with a bright future.

And the Seven Wonders of the Ancient World? There's speculation that the Hanging Gardens of Babylon included fish!

TO LEARN MORE

Aquaponic Gardening: A Step-by-Step Guide to Raising Vegetables and Fish Together, by Sylvia Bernstein. New Society Publishers, Victoria, BC, 2011.

For a research perspective, read "Small-scale Aquaponic Food Production: Integrated Fish and Plant Farming" by the United Nations Food and Agriculture Organization. Fisheries and Aquaculture technical research paper, no. 589, 2014.

Black Gold

Supercharged charcoal—significant sequestration potential

For something that looks like a lump of charcoal, biochar certainly has a great press agent.

The subject of books, articles, blog posts, research papers, workshop presentations, conference talks, and various top-ten-ideas-that-will-change-the-world lists, biochar enjoys a reputation that has, so far, exceeded its actual accomplishments. That's too bad. Its potential ability to address a variety of global challenges is indisputably large, as I'll try to explain, but it has yet to scale up significantly—though I bet that's about to change.

Biochar's appeal is threefold. First, as a supercharged form of charcoal, it has the physiological capability to affect many twenty-first-century challenges simultaneously, including reducing greenhouse gas emissions, increasing food security, boosting water cycles, improving waste management, and assisting renewable energy production. Second, it's a technology, albeit a sooty one, which means it's attractive to the scientific, entrepreneurial, and techno-geek aspects of our society—which partly explains its media charm. It also appeals to the "backyard innovator" in our human nature, as I've seen in various workshops. Third, it's an ancient agricultural practice, which tempts the farmer in us. As the prehistoric tribes of the Amazon Basin knew, this type of "black gold" could elevate soil fertility tremendously.

The trouble is, all of these positive attributes have combined to create a kind of identity crisis for biochar, which I believe is one of the reasons it has struggled to take off. Is biochar a lite form of geo-engineering, a repurposing of indigenous knowledge, or a commercial opportunity for savvy businesses? Or all three? Even its press agent seems confused at times.

What is this black gold exactly?

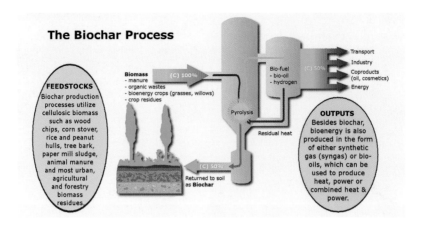

The biochar process, from feedstock to product. *Image by Johannes Lehmann, courtesy of the International Biochar Initiative*

Biochar is produced when organic material, generally plant matter or manure, is heated to very high temperatures in a zero or near-zero oxygen environment, which bakes the carbon into a light but solid structure riddled with millions of tiny holes. The process is called pyrolysis. In nature, it occurs when trees are carbonized by intensely hot forest fires or when wood is engulfed by volcanic lava. In human hands, it usually takes place in a specially constructed oven where temperatures can reach 932°F (500°C) or hotter. In this tightly controlled environment, between 30 and 50 percent of the original carbon is transformed into highly stable biochar. The rest becomes bio-oil and syngas, both of which are exciting to renewable energy experts as potential substitutes for petroleum.

Biochar's appeal as a way to mitigate climate change is straightforward: by baking carbon into a substance that can last thousands of years, we interrupt the natural cycle of decomposition and respiration in which microbes digest organic material and then "burp" CO_2 into the atmosphere. This respiration process is an important source of this long-lived greenhouse gas, and so if we can "lock up" large amounts of carbon as biochar rather than let it decompose, we can (potentially) make a big dent in the blanket of carbon dioxide surrounding our planet. A *really* big dent.

Johannes Lehmann, a professor at Cornell University, recently calculated that if biochar were added to the soil of only 10 percent of the world's farms, nearly 30 *billion* tons of CO_2 would be sequestered—approximately the total amount of humanity's annual greenhouse gas emissions.

That's juicy stuff for any press release.

The waste-management appeal is also straightforward: biochar can be made from a wide variety of biological or "green" waste, including

Eight different biochars made from various types of wood and nut shells, with corn seeds for scale. *Photo courtesy of Josiah Hunt*

lawn clippings, hedge and tree trimmings, and leftover food that would otherwise end up in landfills. Ditto with dairy and horse manure. This is important because landfills and manure lagoons are major sources of methane, a short-lived but potent greenhouse gas. As a bonus, diverting these sources into biochar will reduce vexing waste-disposal challenges.

Of course, composting is another way to put green waste to work regeneratively, but unlike biochar, which is inert, compost is biologically active. Its microbes are busy burping CO_2 into the atmosphere. One intriguing idea proposed by advocates is to mix biochar—which can take the shape of sticks, pellets, or dust—into compost piles. In this way, biochar provides structural stability while the compost provides active biology. This is appealing to farmers, who could add the mix to their soils, boosting its fertility and water-holding capacity. The millions of tiny holes in a piece of biochar provide housing for microcritters, which move in and begin doing their soil-building thing. These holes also wick water from the soil into the biochar (as much as six times its weight) and release the water slowly, supplying the microbes and retarding evapotranspiration, both of which are useful in a drought.

There are other good reasons to like biochar: its stability and resistance to decay enables soil to better withstand flooding and other forms of erosion; it is alkaline by nature, so adding it to acidic soils can help balance pH; and it can help restore carbon-depleted soil in degraded landscapes.

Little wonder, therefore, that prehistoric peoples in South America spent eight thousand years layering the thin, acidic, nutrient-poor soils of the Amazon Basin with a type of biochar called *terra preta*.

At first archaeologists were bewildered by evidence of large human populations in a region that had such poor soil—until they discovered large deposits of terra preta underground. Tests revealed that this homemade version of biochar supplied *exactly* what the soil needed to grow food.

Given all of these impressive benefits, it's fair to ask why biochar hasn't been put to widespread use in our time. One answer: biochar is more complicated than it first appears.

For starters, there are a bewildering variety of biochar types to choose from—225 and counting. There is also a confusing selection of high-tech ovens to bake them in and many (competing) schools of thought about how to produce biochar properly. Then there are technical issues involving thermal physics, feedstocks, and disposal of the bio-oil and syngas produced by the baking process as by-products. There are also practical issues involving transportation and appropriate farming practices, as well as philosophical issues involving competition with compost projects, how to work at scale, and even proper baking temperatures (higher temps produce more stable carbon storage but also use more energy and produce more waste). Finally, there are ethical issues, including the specter of ecologically destructive, industrial-scale biochar plantations.

And then there are the economic hurdles.

Biochar has not yet been produced commercially at a price that makes it competitive with conventional fertilizers or other soil amendments. This could change with the creation of a viable carbon marketplace, where biochar could become a way for polluters to earn "credits" to offset their production of greenhouse gases. Until then, biochar remains mostly in a research-and-development phase. It won't last long, I bet. Biochar has too many important benefits to continue to be underutilized, especially as twenty-first-century challenges mount.

In fact, it has already come a long way in a short time—before 2008, the word *biochar* didn't even exist!

TO LEARN MORE

For more information about the International Biochar Initiative, see: www.biochar-international.org

For the US Biochar Initiative, see: www.biochar-us.org

Night Soil 2.0

New technology resurrecting a forbidden practice

What if we saw human waste as a potential resource instead of a persistent headache?

We know that creating compost and adding it to soils is a quick and efficient way to increase plant productivity and improve carbon stocks underground. What many of us may not know is that the ancient though unsanitary practice of using human waste to return nutrients to the soil is poised to make a comeback—but without the health risks, thanks to a high-tech invention.

While green waste (farm by-products, food leftovers, yard trimmings, and so forth) is an important source of compost today, another kind of waste has a long history as a convenient and effective fertilizer: human feces and urine. It's called "night soil," a euphemism for sanitary waste often collected at night from cesspools, privies, and other stinky sources. It was a common practice around the world for centuries. The ancient Greeks used human feces to fertilize their fields. In Tudor England, night-soil workers were called "gong farmers" and "midnight mechanics." In India, they were included in the caste called "untouchables" and shunned as outcasts (a prejudice that is now outlawed). In the modern era, night soil is still used in a few remote communities.

The main trouble with using feces as fertilizer, besides the smell, is the health risk. Various dangerous bacterial pathogens and parasitic worms live in feces and can easily infect humans unless the raw material is properly processed. This is why communities around the world quickly abolished the practice of using night soil as soon as indoor plumbing and sewage-disposal systems became available. As a consequence, however, a vast (and growing) source of fertilizer is literally wasted—chemically treated and washed downstream or out to sea.

This waste of a potential fertilizer is also a waste of water. In the twentieth century, worldwide water consumption increased sixfold, with a significant portion dedicated to the transport and treatment of human waste. Despite this growth, however, the World Health Organization says 2.5 billion people still do not have access to adequate sanitation. Moreover, using fresh, potable water to move sewage along, as nearly all sanitation systems do, is becoming increasingly unsustainable as the "new normals" of drought and other climate challenges expand globally. Then there's the huge amount of fossil-fuel-generated electricity needed to do the pumping and moving of so much water and waste. The California State Water Project, for example, is the largest single user of energy in the state, consuming more than five billion kilowatt hours per year. For all of these reasons, there is a rising sense of urgency to develop sanitation-disposal methods that dramatically reduce water waste—or eliminate it altogether.

Which brings us back to night soil.

It's back in the news thanks to recent advances in the science of thermophilic composting, including a high-tech pathogenic analyzer called the PhyloChip (officially called a DNA Microarray for Rapid Profiling of Microbial Populations). Developed by Dr. Gary Andersen at the US Department of Energy's Lawrence Berkeley National Laboratory in California, the PhyloChip can simultaneously analyze tens of thousands of microbes in a single water drop or a tiny soil sample. For the purposes of composting, it can track the decomposition of pathogens and indicate the precise moment when the waste becomes disease free.

The key is heat. Lots of heat. Temperatures in a compost pile rise rapidly as microcritters digest the green waste (heat is a by-product of microbial activity and the amount depends on the pile's moisture content, aeration, and carbon/nitrogen ratio). If the compost gets hot enough, all bacteria except the thermophilic (heat-loving) ones will be cooked to death. Since thermophilic bacteria are the good guys (they take care of the smell too), the goal of the PhyloChip is to tell the composter when this process is complete and the soil is good to go for food production. Before the PhyloChip's invention, this process was too risky to apply to human waste—but not anymore!

Enter the Thermopile Project, a program of the Carbon Cycle Institute (CCI), which is based in Marin County, California.

The purpose of the Thermopile Project is both broad and narrow—to advance safe and sustainable solutions to global sanitation, water, and climate challenges, and figure out how to turn human waste into useful compost. In its pilot phase, the Thermopile Project had human waste transported from a nearby national park to a demonstration site on a small ranch near Nicasio, California, where it was combined with woodchips and other vegetative material and then piled high to cook into compost. Heat sensors tracked rising temperatures and a PhyloChip analyzed the bacterial content of the pile. When the cooking was complete and the bad bacteria had been killed off, the compost was ready to be used as organic fertilizer.

These compost piles, which include human waste (well covered by woodchips), on a ranch near Nicasio, California, are part of a pilot project designed to find safe solutions to global sanitation, water, and climate challenges. *Photo courtesy of the Thermopile Project*

Presto! Night soil 2.0.

The practical implications of this innovation are huge. For example, large amounts of night-soil compost could be spread across California's extensive rangelands, triggering soil-carbon sequestration and potentially offsetting significant amounts of carbon dioxide emissions from industrial sources. This is not a guess. A previous science experiment on the same site near Nicasio proved that a half-inch application of compost could boost the carbon cycle significantly and sequester more carbon underground than was respired back into the atmosphere. It has also been demonstrated that adding compost to rangelands increases the water-holding capacity of their soils by as much as 15 to 25 percent. And since California has a large population that produces lots of human waste, it would be the ideal place to give rangeland fertilization with composted night soil a trial on a big scale.

There are other good reasons to consider night-soil remedies. Take energy and water, for example. According to the CCI, California's sewer systems account for 19 percent of the state's total electricity expenditure, 30 percent of its natural gas consumption, and 30 percent of its household water use. Reducing the use of any one of these (or better yet all three) via composting would be a win-win for greenhouse gas emission reduction.

Other benefits include: avoiding damage to aquatic environments from sewage-based nutrients, pharmaceuticals, and endocrine disruptors, especially during floods; reduced reliance on chemical fertilizers in food production; less costly installation, maintenance, and upgrade of sanitation systems, which is especially important in developing

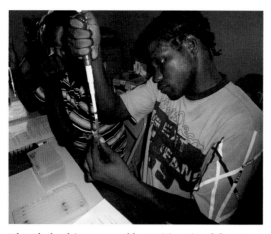

The PhyloChip was used by Haitians in a laboratory to test for pathogens following the devastating 2010 earthquake. *Photo courtesy of the Thermopile Project*

countries; and reduced risk of waterborne diseases in poor and remote parts of the world.

These are the reasons why the PhyloChip was deployed to Haiti after that nation's devastating earthquake in 2010.

For all its hopeful potential, however, challenges to using human waste as compost remain, including the question of how to remove or neutralize "compounds of concern" in the waste, such as pharmaceuticals. Another challenge involves economics—how to make this kind of composting attractive to cities, municipalities, and counties, not to mention entrepreneurs and other businesspeople. Fortunately, many organizations are hard at work on answers to these challenges. It's likely that California's ongoing water crisis, in the form of reduced annual precipitation combined with ever-rising population, will provoke innovation in sanitation management.

Whatever happens ultimately, the Thermopile Project is a great example of innovation on two fronts—using serious science to bring back a long-discarded regenerative practice.

TO LEARN MORE

For more information on the Thermopile Project, visit: www.thermopileproject.com

For more information about the PhyloChip and the Earth Sciences Division of the Department of Energy's Berkeley Lab, visit: http://esd.lbl.gov /research/facilities/andersenlab/phylochip.html

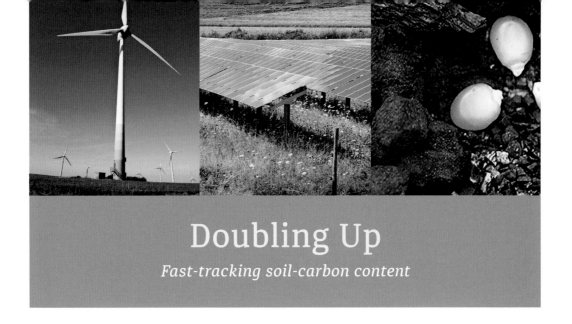

Doubling Up
Fast-tracking soil-carbon content

Can the carbon content of soil be doubled in less than ten years?

It's an important question because in many landscapes around the world a great deal of soil carbon has been lost over the decades as a consequence of tilling, overgrazing, and other unsustainable land management practices that erode the soil. This means the potential for replacing all of this lost carbon through regenerative practices is large, much like refilling a glass half empty of water. That's exactly what has happened on the 500-acre McEvoy Olive Ranch, located near Petaluma, in northern California. And there's more—not only is the farm reaping the many benefits that come with doubling its carbon, including increased soil fertility and water-holding capacity, it has reduced its carbon footprint by installing a wind turbine on the property.

For a tiny place, it has a big story to tell!

It begins in the mid-1800s when Swiss Italian immigrants settled in the area and opened small-scale dairies among the native hardwood rangelands. In the early years, many of the abundant oaks on the property were harvested for firewood to help meet the growing demand for fuel in nearby San Francisco. Although very little of the farm was actually tilled, due predominantly to the steep terrain, hay was grown on the more level meadows. These and other activities reduced the soil-carbon stocks of the land by half.

However, when Nan T. McEvoy purchased the ranch in 1991, the bountiful water, extensive stands of native perennial grasses, and mature woodlands that characterize the landscape were still in good condition. An Italian cuisine maven, Mrs. McEvoy decided to discontinue the livestock production and instead produce one of the finest olive oils in the world. With a commitment not to remove any of the remaining trees on the property, she planted an olive orchard on about 80 acres of the less steep areas of the ranch.

In 1997, Jeffrey Creque was hired to address the question of what to do with the waste products from the ranch's brand new olive oil mill, which he set out to answer by implementing a comprehensive composting program. With a PhD in agroecology and decades of experience as an organic farmer, Creque wanted to help Mrs. McEvoy accomplish her dream and achieve a goal of his own: double the carbon content of the soil from 2 percent (the level in 1997) to 4 percent—his estimate of the amount of carbon that existed in the soil prior to the arrival of European colonists. A full restoration, in other words, of the amount that would have been there naturally.

To accomplish this ambitious goal, Creque and his coworkers embarked on a multifaceted soil-building strategy: apply lots of compost, made from on-ranch olive mill waste, livestock manure, and landscaping debris; employ no-till cultivation, made possible by the maintenance of a permanent cover crop beneath the olive trees; graze sheep in the orchard in a progressive manner; and restore the health of the riparian areas on the property in order to eliminate downcutting gullies.

Creque told me that only 15 to 20 percent of an olive is actually oil, the remainder is water and solids. Historically, in the Mediterranean region this organic material would either accumulate at the milling site or be dumped into a nearby river or the sea. This age-old practice was finally banned in Europe during the 1970s, and today the disposal of olive-mill waste remains a challenge for olive oil producers. Creque's idea for the McEvoy operation was simple: compost all of the green waste and apply it to the soil of the olive orchards, increasing their fertility.

Jeffrey Creque inspects olive-mill-waste compost piles at McEvoy Ranch. Use of the compost doubled carbon content of the soil in less than ten years. *Photo by Courtney White*

In this way, a problem became a solution.

"Olive oil is produced from the current season's photosynthetically derived carbon," Creque said. "If the ranch exports only oil, it essentially removes nothing permanently from the soil. By avoiding tillage and returning all residuals to the land, the olive oil agroecosystem takes in more carbon from the atmosphere than it emits. Done well, olive oil production can be an essentially permanent, regenerative form of agriculture."

Data supports Creque's claims. Dozens of soil samples are taken every year from all over the ranch and sent to a laboratory for analysis. While results have shown year-to-year fluctuations in the organic-matter content of the soil, mostly due to weather and sampling variables, the trend over time has been clear: upward. In fact, after 10 years the carbon content in all samples has begun to hover around 4 percent. This means that the olive ranch is sequestering more CO_2 than it did back in 1997. It's also more productive and its soils are holding more water. Goal accomplished!

Creque didn't want to stop there, however. By tackling the restoration of the ranch's riparian areas, a new challenge emerged, along with a new carbon-sequestration opportunity: managing surplus riparian vegetation (especially willows) for compost production. As the overall productivity of the ranch increased, the volume of carbon sequestered in standing biomass and soils, and potentially available for composting, also increased. So why stop at 4 percent soil carbon?

"There's no reason to think that we can't increase soil carbon in our agricultural systems to levels above those that would occur without management," Creque told me. "Besides, there are no downsides to trying and lots of upsides, especially for agricultural productivity, sustainability, and climate change mitigation. If we can manage our soils to store more carbon, we'll also enable them to store more water, while reducing the volume of CO_2 in the atmosphere. That's a big upside."

Creque notes that across the nation, millions of tons of organic waste—food, grass clippings, branches, manures—go into landfills every year, where they produce a lot of methane, a potent greenhouse gas. Why not compost them instead, he said, and then spread the compost across farms and rangelands where it could provide multiple benefits to the landowner and the public? Of course, there's a financial and a carbon cost to hauling this material around, but it could be offset by increased ecological productivity and potential carbon credits, not to mention benefits to the Earth's climate system. This would be an economical way to reduce emissions on one hand and increase mitigation on the other.

A wind turbine reduces the ranch's greenhouse gas emissions while meeting about half of its electrical energy needs. *Photo by Courtney White*

Next on the agenda was the wind turbine.

In 2009, in an effort to produce renewable energy and reduce its carbon footprint, McEvoy Ranch installed a 225-kilowatt wind turbine on a hill overlooking the orchards. It is estimated that the turbine reduces the ranch's greenhouse gas emissions by 110 tons of CO_2 each year, while meeting about half of its electrical energy needs (solar panels provide much of the rest of the ranch's energy). However, at the time it was the largest wind project in Marin County, and the process of getting it built, including various regulatory hurdles, became a challenge. For example, the original height of the turbine was scaled down from 248 feet to 98 feet by orders of the county commission after neighbors and others complained about its visual impact. As a result of the debate, the county imposed restrictions on all future wind projects.

"Increasing soil carbon is relatively easy," Creque told me with a wry smile. "Overcoming the bureaucratic challenges to installing sustainable energy systems can be much more difficult."

Nevertheless, as the McEvoy Ranch demonstrates, problems can become solutions when attitudes shift and appropriate technologies are applied—and that's a lot to cheer about!

TO LEARN MORE

For more information on the McEvoy Ranch,
see: www.mcevoyranch.com

"Regenerative Organic Agriculture and Climate
Change: A Down-to-Earth Solution to Global Warming," a
white paper from the Rodale Institute, 2014.
Available through www.rodaleinstitute.org

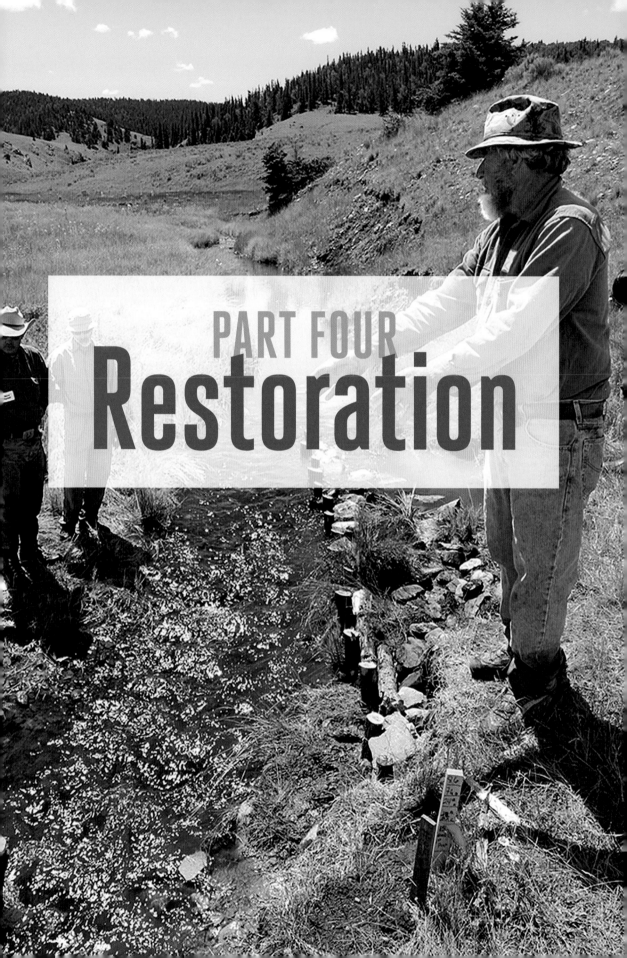

PART FOUR
Restoration

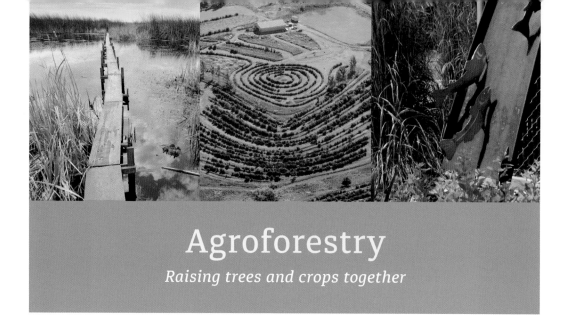

Agroforestry

Raising trees and crops together

Her do you like your trees—as poetry, or science? Or both?

A recent report by scientists at Oxford University concluded that the best "technology" (their word) available to fight climate change right now, besides building soil carbon, is to plant a tree. By their nature trees soak up CO_2 and store it for long periods of time, if they are left alone to grow and don't burn up in a forest fire. In mind-numbing jargon, however, the report scientists declared trees to be a "Negative Emissions Technology," which means they have the potential to remove and sequester significant amounts of CO_2.

Personally, I look at trees less analytically, preferring poet Joyce Kilmer's classic perspective *"I think that I shall never see / A poem lovely as a tree."*

Nevertheless, during the field-tour portion of an international conference on resilience, which took place in southern France, I decided to take a look at the science of trees, specifically the intercropping of trees and crops. Having lived nearly all of my life in the arid American West, my interaction with trees had mostly been confined to sweeping expanses of ponderosa pine, spruce fir, aspen, piñon, and juniper trees. Beyond the fruit and nut orchards that I had seen in California's Central Valley, the practice of raising trees and crops together was a foreign concept to me.

Conference organizers bussed our group to a research site called Restinclières, managed by the National Agricultural Research Institute (INRA), whose local headquarters was housed in a lovely eighteenth-century chateau (ah, France). Under a blazing sun, we strolled into a field of wheat growing amid a grove of young, evenly spaced walnut trees and listened to a scientist explain the environmental and economic benefits of intentionally intercropping an orchard with farm

Wheat growing between walnut trees at the Restinclières research site in France. Agroforestry is the intentional integration of trees and crops.
Photo by Courtney White

crops. In Europe, he told us, open space is at a premium, so any system that integrates two sources of food or fiber on one plot of land is worth implementing and studying, as they were doing at Restinclières.

Broadly, this practice is called agroforestry—and it's not something new. Variations on tree and plant intercropping have been practiced for centuries, including:

- Silvopasture, in which fruit, nut, or timber trees are intercropped with pasture, providing shade and shelter for grazing livestock;
- Alley cropping, in which vegetables, grains, flowers, herbs, or bioenergy feedstocks are planted between tree rows;
- Multistory cropping, in which food-producing shrubs and trees of varying heights are grown together;
- Forest buffers along rivers and streams, which stabilize the river banks while providing a source of food;
- Windbreaks, shelterbelts, hedgerows, and living fences, all of which can support agricultural activities while providing protection to wildlife and landowners.

No matter which practice is selected, the key to successful agroforestry is choosing trees and plants that have complementary ecological relationships—walnut trees and wheat, for example. Often, these relationships are embedded in local knowledge and ages-long experience, but understanding how they work scientifically and exploring possible new relationships has become an important component of agricultural research, as we witnessed that day under the hot sun.

According to the USDA, there are four "I's" to agroforestry: each system must be *integrated*, *intensive*, *intentional*, and *interactive*. In other words, agroforestry plots need to be carefully planned, managed, and monitored. The plots can range from highly cultivated to semiwild, but in all cases the four "I's" rule.

When agroforestry is done successfully, water quality and pollinator habitat are improved. The wide spacing of trees accelerates their growth, and the quality of wood is improved due to reduced competition and thinning. Trees protect animals from extreme weather (heat, wind), so grazing seasons can be extended. Keeping the ground surface covered by plants can suppress weeds, protect the soil, reduce erosion, and help restore degraded land. And multiple crops provide a diversified income source for landowners and can mitigate risks resulting from price spikes, disease outbreaks, and floods.

The list of benefits goes on. Basically, agroforestry is good for soil microbes, plant roots, livestock, wild animals, humans, and economies. Oh, and the climate benefits too, thanks to the ability of trees and plants to soak up a great deal of CO_2. Amazingly, however, despite all these benefits agroforestry is not as widely implemented as one might expect. In 2012, for example, the US Department of Agriculture included for the *first time* a question about agroforestry in its comprehensive Census of Agriculture. I suspect the main reason for this situation is the dominance of "silo" thinking in agriculture, by which enterprises such as sheep and cattle are kept tidily separate from one another. But as science is beginning to clearly demonstrate, there's a lot to be gained from breaking down these silos and mixing things up—just as nature does.

None of this should be news. Agroforestry practices have been used by indigenous peoples around the world for centuries to produce medicine, fuel, food, clothes, sacred objects, living quarters, and much more. Early European settlers in the New World adopted similar practices, weaving livestock into the mix. However, the rise of industrial agriculture in the twentieth century, especially the advent of tractor farming and applications of chemical fertilizer, ended most of these age-old integrated approaches to land management. Making matters worse, the adoption of industrial practices was spurred along by the separation of scientific research into specialized silos of its own, resulting in a reductionistic (antiholistic) approach to food production and ecosystem health.

Fortunately, integrated approaches like agroforestry did not disappear altogether. They were conserved in pockets around the US by native peoples, organic farmers, and others. In the 1990s, the number of these practitioners began to grow, as concerns about sustainability and other challenges spread. It took a while but

In a silvopasture system, trees are intercropped with pasture, providing shade and shelter for grazing livestock. *Photo courtesy of New Forest Farm*

eventually the research community began to focus on these practices, and this has developed into a robust scientific effort, as I saw in France.

According to the scientist who led our field tour at Restinclières, important areas of research in agroforestry systems today include their ability to build ecological resiliency in the face of adverse conditions brought on by climate change; measuring precisely their capacity to sequester atmospheric carbon; and which plant and tree species are the most complementary with one another. While the amount of forested land is decreasing globally due to habitat conversion and destruction, he told us, the numbers of trees on farms is rising. Add in the food, fuel, and timber value of trees and you create a win-win-win scenario for all involved. Science can help—though not if we keep thinking of a tree as "technology," I think.

At the end of the tour, as we walked along a pleasant path back to INRA's headquarters in the chateau, I reflected once again on the elegance of natural processes and the inelegant ways we humans keep messing up nature's good ideas. We've been especially hard on trees over the centuries. Fortunately, trees grow back, especially if we give them half a chance. That's good news on many levels. However discouraging our behavior might be, it is heartening to see old practices revived with modern twists, especially in long-used landscapes like those in southern France. Besides, there's just *something* about trees. They stir primordial emotions in us, often driving us to flights of literary fancy. *"Poems are made by fools like me,"* wrote Kilmer, *"but only God can make a tree."*

Will we learn from our mistakes? Can we repair the damage we've done? After a day in the hot sun, it certainly felt like a possibility!

TO LEARN MORE

For a USDA guide on agroforestry, see:
www.usda.gov/agroforestry.html

For an INRA site focused on agroforestry
(with slideshow), see: http://www1.montpellier.inra.fr
/safe/english/agroforestry.php

Restoration Agriculture

Healing damaged land while feeding people

I bet Aldo Leopold would have admired Mark Shepard.

In the 1930s, renowned conservationist Aldo Leopold and his family purchased a worn-out patch of Wisconsin farmland, beaten and neglected by previous owners, and set out to restore it to ecological health, principally by planting trees. Lots of trees. In the 1990s, Shepard set out on much the same quest less than fifty miles to the west. Unlike Leopold, however, who selected pine, cherry, and oak saplings for purely ecological purposes, Shepard designed the repair work on his land with the goal of producing staple food crops from trees, including hazelnuts, chestnuts, pine nuts, and apples. It was a deliberate attempt to demonstrate that Leopold's idea of ecological restoration could simultaneously heal damaged land and feed people sustainably.

Shepard coined the term "restoration agriculture" to describe this effort.

But let's back up for a second and revisit two quotes by Aldo Leopold that laid the foundation not only for Shepard's work but for restoration efforts in general.

> "Land . . . is not merely soil; it is a fountain of energy flowing through a circuit of soils, plants, and animals. Food chains are the living channels which conduct energy upward; death and decay return it to the soil. The circuit is not closed; some energy is dissipated in decay, some is added by absorption from the air, some is stored in soils, peats, and long-lived forests; but it is a sustained circuit, like a slowly augmented revolving fund of life."
> —Aldo Leopold (A Biotic View of Land, *1939*)

"A land ethic . . . reflects the existence of an ecological conscience, and this in turn reflects a conviction of individual responsibility for the health of the land. Health is the capacity of the land for self-renewal. Conservation is our effort to understand and preserve this capacity."
—*Aldo Leopold* (A Sand County Almanac, *1949*)

Leopold focused much of his ecological conscience and legendary energy on environmental objectives, including restoring native prairie and urging farmers to adopt conservation practices that created healthy habitat for game and other types of wildlife. The aim of restoration agriculture is broader: the deliberate design of farms patterned on natural ecosystems. *A Sand County Almanac* meets permaculture, in other words. For Shepard, the ultimate goal is to create a deeply diverse agroecosystem, full of beneficial synergies that develop richness and complexity over time—that also produce food to eat.

Just like a forest.

That's why Shepard and his family chose the name New Forest Farm for their enterprise. Located in Viola, Wisconsin, the farm's 106 acres harbored only 100 trees in 1995, the year the Shepards took over. Today there are nearly 250,000! Like Leopold, Shepard aimed to mimic the oak savanna that once dominated the upper Midwest, both in its structure (vertical and spatial) and its mix of tree and shrub species. However, the plants he selected and where he planted them differed substantially from the Leopold family project. Shepard aimed to replace the annual cropping system (corn, soybeans) that had destroyed the health of the land with a regenerative, perennial ecosystem that could provide food and profit for himself and his family.

This aerial view of New Forest Farm, near Viola, Wisconsin, shows its deliberate design patterned on natural ecosystems. *Photo courtesy of New Forest Farm*

"What I'm not talking about is doing purist restoration," Shepard said in an interview in *Acres* magazine. "We've designed a savanna system here with the primary goal of providing staple food crops. Because it's a rich, diverse ecosystem, it also supplies all kinds of everything else."

Shepard credits his roots in central Massachusetts and a poisoned river for the unlikely journey that led to New Forest

Farm. The Nashua River, which flowed in a big loop around his childhood home, could be red, green, orange, or cobalt blue, depending on what the upstream paper mill had most recently discharged into the water. When the pollution was eventually abolished by law, Shepard was amazed at how quickly the river came back to life, sprouting vegetation and filling with fish. The healing power of nature, he realized, was a powerful thing.

Meanwhile, his parents had begun to chop wood and grow a large garden in an earnest attempt to make the family more self-sufficient. This also had a big effect on Shepard, who not only came to prefer the cool woods to the hot sun of the garden but wanted to grow food staples, not just fruits and veggies.

"Nature looked after it all," he said. "The woods provided our family with a considerable amount of food, apparently free of charge and for no more work than was required to harvest and preserve it."

After college, Shepard and his wife homesteaded in Alaska for eight years, which made them think long and hard about food, trees, and survival. By a serendipitous accident, they wound up purchasing an exhausted, hilly farm in western Wisconsin.

Because Shepard considers water to be the most important nutrient on the planet, he began his restoration effort by designing a water-capture system called a Keyline with the goal of storing 100 percent of the rainwater that fell on the farm. He built hundreds of swales and ponds, and continues to add more every year. In the process, he discovered that he can build soil from the "bottom up" (more water increases microbial life, which naturally converts the subsoil into topsoil and literally fluffs it up), as well as relying on the usual methods that build soil from the top down (the accumulation and decay of litter and other organic matter on the soil surface).

He discovered, in other words, how to expand Aldo Leopold's idea of a "revolving fund of life."

The next step in restoration agriculture is to "know your biome," as Shepard puts it. What would nature do here? Identify which plants, animals, soils, and rainfall patterns characterize your region. In particular, target ecological succession pathways—the local, predictable, and orderly progression of plants (weeds to grasses to shrubs to trees) that nature wants to see occur over time. From this suite of plants, a farmer can choose which ones have the best likelihood of producing profitable yields. Once you get the plant mix right, which can take a certain amount of trial and error over time, Shepard advises, then you add livestock and pollinators. The goals are jump starting the soil food web and building carbon stocks underground.

Hazelnut seedlings in tree shelters. New Forest Farm's goal is to simultaneously heal damaged land and feed people sustainably. *Photo courtesy of New Forest Farm*

In all of this, Shepard believes trees have important advantages over annual crops. Photosynthesizing starts much earlier in the year with trees and lasts longer into the fall; because of their height and width, they capture more sunlight and other resources; they don't have to build their bodies from scratch every growing season; they grow bigger each year; and they live a long time. And they are much less labor intensive.

"One of the wonderful things is I don't have to plant my crops ever again," Shepard said. "The trees just keep growing and the animals kind of take care of themselves."

Today, New Forest Farm produces chestnuts, hazelnuts, pine nuts, walnuts, hickory, apples, cherries, asparagus, and winter squash. They produce hard cider and raise cattle, pigs, lambs, turkeys, and chickens, which are grazed in a manner that mimics the behavior of wild herbivores. "Our animals stay healthy and happy," the Shepard family writes on their website, "eating an incredible diversity of nutritious and medicinal forage and are all treated with dignity and respect."

It all starts and ends with nature.

"The ecological system has been here on this continent for a bazillion years and it's done just fine," Shepard said in the *Acres* magazine article. "We need to convert our annual crop farms into perennial ecosystems. One tree at a time."

Words that would have warmed Aldo Leopold's heart.

TO LEARN MORE

For the New Forest Farm website, see:
www.newforestfarm.net

Restoration Agriculture: Real-World Permaculture for Farmers, by Mark Shepard. Acres
USA Press, Austin, TX, 2013.

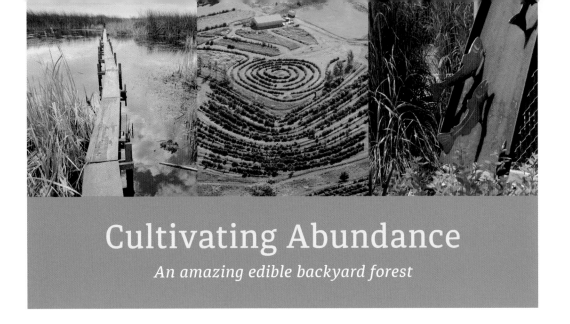

Cultivating Abundance

An amazing edible backyard forest

This is a story about two plant geeks, an urban sweet spot, and edible forests.

The two self-described plant geeks are Eric Toensmeier and Jonathan Bates. The edible forest garden they planted in 2005 resides on one-tenth of an acre behind a duplex home they bought in the Rust Belt city of Holyoke, Massachusetts. Although tiny, the property had big problems: the backyard was lifeless, the soil full of brick and concrete bits, the narrow alleyways in deep shade; the steep, short front yard was covered in asphalt; and the legal terrain was hostile to composting, water harvesting, and livestock, even chickens.

It was perfect, in other words.

That's because Toensmeier and Bates wanted to see if they could bring a tiny spot of badly damaged land back to health by creating an edible ecosystem on it. That meant a forest garden, which is defined as an ecologically designed community of mutually beneficial perennial plants intended for human food production. Think fruits, nuts, berries, and certain veggies. Could they bring lifeless land back to life by gardening every square inch, they asked, creating a diverse and edible landscape? Would permaculture strategies developed in Australia work in the Northeast US? Could they grow banana plants in wintry western Massachusetts? If so, what else could they grow and how could their project serve as a role model for ecological restoration in cities using native perennial plants?

The plant geeks set out to find out.

The route to Holyoke began in 1990 when Toensmeier became intrigued by permaculture and its basic equation: indigenous land management knowledge + ecological design + sustainable practices = landscapes that are more than the sum of their parts. He was particularly excited by its utility for designing food-producing ecosystems—but would it work in his native New England? The answer arrived when Toensmeier met Dave Jacke, an expert in edible ecosystems. Jacke defines an edible forest garden as a perennial polyculture of multipurpose plants on a small plot of land that provide what he calls the seven "F's": food, fuel, fiber, fodder, fertilizer, "farmaceuticals," and fun. It's a forest, in other words, except it's a garden. He means it is gardening *like* a forest, not *in* a forest. Forest gardeners use the structure and function of a forest as a design strategy while adapting the design to meet human needs.

Back in Holyoke, Bates and Toensmeier knew that one advantage to perennial plants, besides providing tasty food, was their ability to build soil, control erosion, improve rainfall capture, and sequester carbon. These could be very useful qualities in a blighted urban context, they thought. And there was another important advantage to perennials—minimal maintenance.

"Having worked on annual vegetable operations and experienced the hard labor of planting and caring for annuals," Toensmeier writes in his and Bates's book *Paradise Lot*, "I considered low-maintenance edible perennial vegetables an appealing alternative."

The key to an edible ecosystem is a design that is as multifunctional as possible. To come up with one, Toensmeier and Bates moved into the duplex in January 2004 and then spent an entire year observing and analyzing their one-tenth acre and contemplating their design. What part of the property received the most sunlight year-round (for the greenhouse)? Where was the best place for the pond? What guilds of plants would work best together in which part of the backyard?

Scouting around the neighborhood for an ecological role model, they were delighted to discover a "feral landscape" behind an old shopping center. It was ten acres of shrubs and wildflower meadows, just right for their purposes. Nature was well on its way to healing the two-decades-old scar created by the development, and by studying the plants, Toensmeier and Bates gained valuable clues as to what nature likes to grow in a disturbed urban ecosystem.

"Most gardeners would not be excited about the species that were growing in the abandoned area behind the shopping center," Toensmeier wrote, "But to me, any plant community that can grow in such terrible conditions is a welcome one."

In 2005, after sheet mulching the bare ground behind the duplex (layers of straw, compost, organic fertilizers, and cardboard) they planted native persimmon, pawpaw, beach plum, clove currant, blueberries, chinquapins (bush chestnuts), hog peanuts, grapes, pears, and the nonnative kiwifruit (carrots and apples are also nonnative, Toensmeier notes). In the front yard they planted banana trees.

The condition of the duplex backyard in 2004, before the edible forest was planted by Toensmeier and Bates. *Photo by Eric Toensmeier*

The same duplex backyard seven years later. Crops include native persimmon, pawpaw, beach plum, clove currant, blueberries, grapes, and pears. *Photo by Eric Toensmeier*

By 2007, the garden was coming to life, a consequence of improving soils and the attractive habitat they had created for beneficial insects. The shrubs, perennials, and young trees were doing well, and the front yard already looked like a mini tropical paradise. The banana trees, sheltered from westerly winds and collecting heat from the asphalt driveway, their roots protected from winter snows, became showstoppers in the area. Drivers slowed down to gawk. Puerto Rican neighbors asked permission to harvest leaves for tamales.

By 2009, the backyard ecosystem was showing "emergent properties," as the geeks described it: things were happening that were more than the sum of their parts. For example, a blue salamander discovered under a persimmon tree in the garden meant that the edible ecosystem was attracting forest animals to patrol its understory. No salamander could have survived in the yard in 2004.

In 2010, Bates kept a log of the amount and kinds of food coming into the kitchen from the garden. He estimated that over six months, he and Toensmeier harvested four hundred pounds of fruits and vegetables from their one-tenth acre, a total that was bound to rise in subsequent years as the edible ecosystem reached its full capacity. Best of all, the incredible yields were being produced with virtually no labor. It was a testament not only to the success of their design, but to the regenerative power of nature to produce life.

"The abundance in our garden comes to us in a self-renewing way," Bates wrote in a sidebar for *Paradise Lot*. "Our fruit trees are surrounded not by grass and asphalt, but by other useful and edible easy to care for plants. After eight years, with very little care from us, all the plants are providing food, medicine, mulch, fodder, beauty, habitat, knowledge, seeds, and baby plants."

"How is it that the abundance that I am now seeing in the garden," he exclaimed, "and in life, was hidden from me all this time?"

Bates leads a tour. In six months, over four hundred pounds of fruits and vegetables were harvested from their one-tenth acre. *Photo by Eric Toensmeier*

For Toensmeier, their little sweet spot demonstrated that cold-climate forest gardening can work. He and Bates created a multistoried forest garden in Massachusetts that can produce food from trees, shrubs, herbs, and fungi, even in the shade. They showed that ponds can grow food, asphalt can be a boon to tropical plants, and a good time can be had by all. There were challenges and setbacks, of course (detailed in *Paradise Lot*), but after eight years they had accomplished everything on their original to do list, and more.

"While sustainability is focused on maintaining things as they are, regenerative land use actively improves and heals a site and its ecosystems," Toensmeier wrote. "Regenerative agriculture . . . achieves these goals while also meeting human needs. It's kind of an important topic for humanity this century."

What Toensmeier and Bates started in the backyard of their duplex continues to this day with workshops, tours, plantings, and harvestings —which is good news indeed!

TO LEARN MORE

*Paradise Lot: Two Plant Geeks, One-Tenth of an Acre, and
the Making of an Edible Garden Oasis in the City*
by Eric Toensmeier and Jonathan Bates.
Chelsea Green Publishing, White River Junction, VT, 2013.

Eric Toensmeier and Jonathan Bates each have
websites. Toensmeier's is www.perennialsolutions.org
and Bates's is www.foodforestfarm.com

The Paradise Lot blog is at:
www.paradiselotblog.wordpress.com

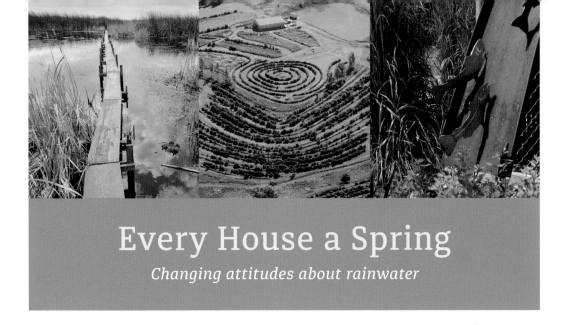

Every House a Spring

Changing attitudes about rainwater

Sometimes innovation isn't a thing, a practice, or a new technology, but simply a different way of looking at the world.

While attending a rainwater-harvesting workshop near my home, I was surprised to hear the instructor say, "Think of every house here as a potential spring." What did he mean? I live in a high, cold desert on a former ranch turned into an exurban subdivision where water scarcity is always a worry. I raised my hand. "There are probably two thousand homes here," I asked, "are you saying they could become two thousand separate springs?" He nodded his head. I had never thought of our community that way before: the Land of Two Thousand Springs! And nothing had changed except my attitude—plus acquiring a little knowledge about rainwater harvesting.

The practice is exactly what it sounds like—a method for capturing, storing, and utilizing water that falls free from the sky. It's an ancient activity, of course, especially in arid environments, but it's largely disappeared from industrialized societies as groundwater pumping, dam building, canal constructing, and modern plumbing made the need for water conservation obsolete. Or so we thought. As drought painfully reminds us, every drop of fresh water is precious. And if the current persistently dry times become something like a new normal—as climate scientists predict—then it's in our interest to harvest every drop we can find.

Enter Brad Lancaster, who lives in water-sparse Tucson, Arizona. In 1994, Lancaster and his brother tried to save a drought-stressed orange tree in the backyard of their home by digging a shallow depression around the tree, adding mulch, and then redirecting rainwater from a neighbor's roof (with permission) into the depression. "The results amazed us," Lancaster wrote in his book *Rainwater Harvesting*. "After a single rain, the

This artistic downspout is a creative way to harvest rainwater—capturing, storing, and utilizing water that falls free from the sky. *Photo courtesy of Brad Lancaster*

tree burst out with new leaves, a dreamy show of fragrant blossoms, and an abundant crop of fruit that was soon converted into tasty marmalade."

Inspired, they planted shade trees around their house, which essentially eliminated the need for their evaporative cooler. Then they redirected stormwater that would normally barrel off the roof and away down the street onto nineteen young trees that they planted in the right-of-way strip in front of their home. Soon the trees were blooming. Next they began recycling water from the house, dropping their daily water use from the Tucson household average of 114 gallons per day to less than 20, all of which caused their water and electricity bills to plummet.

In fact Lancaster was so successful at cutting the water and electricity use in his own home that he was visited by employees of utility companies five separate times to make sure the meters weren't broken!

In making his case for rainwater harvesting, Lancaster begins with what he calls the standard paradox: although a quarter-acre lot in Tucson receives about 67,000 gallons of rainwater a year, nearly all of that pours off roofs, yards, driveways, and parking lots and disappears into storm drains. Meanwhile, the average single-family residence in Tucson uses about 120,000 gallons a year, roughly half of that for outdoor plants, nearly all of which is groundwater pumped up from aquifers or river water delivered long distances by canal. The rain is free, the groundwater is expensive—a financial as well as a cultural paradox.

The answer, says Lancaster, is to change our attitude. Our goal should be to *retain* not drain; *harvest* not waste.

Lancaster's methodology focuses on creating what he calls "living nets" of water-holding vegetation and topsoil around a home, each designed to catch rainwater. His strategy includes removing as much impervious surface material (cement, asphalt) as possible, building bowl-like earthworks to capture runoff, installing tanks to store the water, using streets as irrigators during storm events, installing low-flow appliances and greywater systems (to capture and redirect water

Redirecting stormwater to the root zone of trees can help drop household daily water use significantly, causing water and electricity bills to plummet.
Image courtesy of Brad Lancaster

from baths and showers), and integrating all of these to significantly reduce a household's demand on the municipal water system. And if the "living net" vegetation produces something edible, then you get an added bonus—food!

According to Lancaster, the costs of installing a rainwater-harvesting system can be low, especially if you do the labor yourself. He recommends creating a water budget first, which involves figuring out how much water can be reliably generated on your lot, followed by decisions on the best harvesting strategies for your needs and finances. Generally, what works best is a combination of rainwater-harvesting earthworks, greywater systems (if additional water is needed), and tree planting with good surface mulch. If a vegetable garden is also part of the plan, then you could employ cisterns as well. It's amazing what these simple changes can create, as Lancaster can speak from firsthand experience.

"Our lot was once hot, barren and eroded, with a house that could only be made comfortable by paying to mechanically alter its climate," Lancaster wrote. "Now our yard is an oasis producing 15-25% of our food and after growing trees and installing solar panels to power fans we no longer pay a cent to heat and cool our home."

Which is quite an accomplishment in the Sonoran desert.

"On our 1/8-acre lot and right-of-way we currently harvest about 100,000 gallons of rainwater during an average year of rainfall with 5,000-gallon capacity in tanks, and much more in soil and vegetation," he wrote, "while using less than 20,000 gallons of municipal groundwater per year for our domestic needs and irrigation during dry spells. Four-fifths of the water we consume now comes from our own yard and right-of-way, not the city supply."

Sounds like a spring to me.

Rainwater has other advantages over city water besides being free. It is one of the purest forms of water on the planet. Soft water—unlike mineral-rich, hard groundwater—is perfect for cooking and washing. Also, rainwater is very low in salt and is thus ideal for plants. Lancaster's rainwater-harvesting strategies can help to recharge local wells, springs, aquifers, and streams because they increase the amount of water that infiltrates into the ground rather than running off into a storm drain.

Then there's the big picture: How are humans going to meet the rising demand for fresh water in the twenty-first century? Only 3 percent of all of the water on Earth is fresh, and the hydrological cycle in a particular location can only make so much rainwater available per year—11 inches in Tucson's case. Toss in the double whammy of a growing global population and diminishing rainfall events due to climate change and you have a serious problem, especially in drylands, which make up half of the planet's terrestrial surface. Fortunately, this crisis can be largely alleviated by harvesting rainwater more effectively.

Lancaster sums up the benefits of rainwater harvesting this way: it provides drinking water, generates high-quality irrigation water, supports vegetation that acts as living air conditioners, lowers utility bills, enhances soil fertility, grows food, provides beauty, increases local water resources, reduces demand for groundwater, reduces pollution, boosts wildlife habitat, and endows us with the skill to become sustainable. It is a cost-effective and time-tested way to make your home and community more resilient for the long run.

One spring at a time—it's an attitude we can all share.

TO LEARN MORE

Rainwater Harvesting For Drylands and Beyond,
by Brad Lancaster. Rainsource Press, Tucson, AZ, 2013.

Brad Lancaster's website is at:
www.harvestingrainwater.com

An instructional video featuring
Brad Lancaster can be viewed at:
https://www.youtube.com/watch?v=k9Ku_xpyLK4

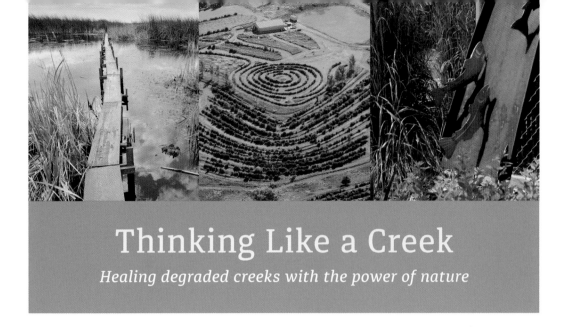

Thinking Like a Creek

Healing degraded creeks with the power of nature

Renew, heal, reaffirm, nurture, rekindle, revitalize, repair, revive, mend, soothe, rebuild, fix, regenerate, reinvigorate—all words to live by today.

I'll explain why with a short story: I once heard about a man who put short fences across a cattle trail in the sandy bottom of a canyon in desert country. The presence of the fences forced cattle in the canyon to meander in an S-pattern as they walked. This, in turn, encouraged stormwater to follow the same meander as it rushed down the wash, which slowed erosion that would have occurred otherwise. It was a simple solution to a persistent problem—and heretical too. That's because the standard fix for degraded creeks is to spend a bunch of money on cement, wire baskets full of rocks, and diesel-driven machines. Putting fences in the way of cattle and letting them do the work? How cool was that?

The man was Bill Zeedyk, a retired biologist with the US Forest Service, now a creek restoration specialist. Was the story true, I asked him? He reassured me that it was. Recognizing that water running down a straight trail will cut a deeper and deeper incision in soft soil with each storm, Zeedyk talked the local ranchers into placing fences at intervals along the trail. This would force the cows to create a meander pattern in the soil precisely where he thought nature would do so in their absence. Water likes to meander, which is nature's way of dissipating energy, and it will gravitate toward doing so even when it's temporarily trapped in a cattle-caused rut (or human-caused hiking trail). Zeedyk's fence idea was a way to speed the process up.

What happened after the fences were put in, I asked? The water table came up as vegetation grew back, replied Zeedyk, because the water was now traveling more slowly and had a chance to percolate into the ground, rather than run off as it had before. Eroded banks began to revegetate as

Bill Zeedyk teaches a restoration workshop. In front of him is a post vane structure, which redirects flowing water away from an eroding bank. *Photo by Courtney White*

the water table rose, and more water appeared in the bottom of the canyon, which encouraged riparian plant growth.

"Nature did all the heavy lifting," Zeedyk said. Then he added, with a knowing smile, "It worked great, until someone stole the fences."

Over the years, Zeedyk has developed a very effective set of low-cost, low-labor techniques that reduce erosion, return degraded riparian (creek) areas to properly functioning conditions, and restore wet meadows to health. This is important because a big part of the world's land exists in an eroded condition, mostly the result of poor land management past and present.

To repair this kind of damage, Zeedyk has put together a toolbox designed to "heal nature with nature." It includes:

- One-rock dams/weirs, grade-control structures composed of wooden pickets or rocks that are literally one-rock high and simulate a "riffle" effect in creeks;
- Baffles/deflectors, wedge-shaped structures that steer water flow;
- Vanes, a row of wooden posts that project upstream to deflect water away from eroding banks;
- Headcut control structures/rock bowls, to slow or stop the relentless march of erosion up a creek and trap water so that vegetation can grow.

Many of these structures are placed directly in a watercourse. Vanes and baffles, for instance, are used to deflect stream flow. Weirs are used to control streambed grade and pool depth. One-rock dams are used to stabilize bed elevation, modify slope gradient, retain moisture, and nurture vegetation.

The goal of all these structures is to stop water from downcutting a creekbed, often by "inducing" an incised stream to return to a

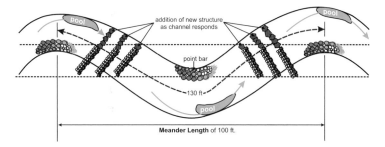

How induced meandering works. The placement of stone-and-stick structures at key points causes the creek to become sinuous again, slowing water
down. *Image by Tamara Gadzia for the Quivira Coalition*

dynamically stable channel through the power of small flood events.
Zeedyk calls this "induced meandering."

When a creek loses its riparian vegetation, including grasses, sedges,
rushes, willows, and other water-loving plants, it tends to straighten out
and cut downward because the speed of water is now greater, causing
the scouring power of sediment to increase. Over time, this downcutting results in the creek becoming entrenched below its original
floodplain, which causes all sorts of ecological havoc, including a drop
in the water table. Eventually the creek will create a new floodplain at
this lower level by "re-meandering" itself, but that's a process that often
takes decades. Zeedyk's idea is to goose the process along by forcing
the creek to re-meander itself as the result of his carefully calculated
and emplaced vanes, baffles, and riffle weirs. Once water begins to slow
down, guess what begins to grow? Willows, sedges, and rushes.

"My aim is to armor eroded stream banks the old fashioned way," said
Zeedyk, "with green, growing plants, not with cement and rock gabions."

The employment of one-rock dams typifies Zeedyk's naturalistic
approach. The conventional response of landowners to eroded and
downcut streams and arroyos has been to build check dams in the
middle of the water course. The idea was to trap sediment behind a
dam, which would give vegetation a place to take root as moisture
is captured and stored. The trouble is that check dams work against
nature's long-term plans.

"All check dams, big or small, are doomed to fail," said Zeedyk.
"That's because nature has a lot more time than we do. As water does

its work, especially during floods, the dam will undercut and eventually collapse, sending all that sediment downstream and making things worse than if you did nothing at all."

"The trick is to think like a creek," he continued. "As someone once told me long ago, creeks don't like to be lakes, even tiny ones. Over time, they'll be creeks again."

Zeedyk's one-rock dams don't collapse because they are only a single rock high. Instead, they slow water down, capture sediment, store a bit of moisture, and give vegetation a place to take root. It does take time, however, which challenges our modern "microwave" culture, which is conditioned to immediate results. "But nature often has the last word," said Zeedyk. "It took one hundred and fifty years to get the land into this condition; it's going to take at least as long to get it repaired."

The key is to learn how to read the landscape and become literate in the language of ecological health.

"All ecological change is a matter of process," said Zeedyk. "I try to learn the process and let nature do the work, but you've got to understand the process first. If you don't, you can't fix the problem."

Over 20 years and across a dozen states, Zeedyk has implemented hundreds of restoration projects, healing miles of riparian areas—all by thinking like a creek. He's been successful not simply because he understands ecological processes or because he's created an effective methodology, but also because his goals and his attitude are positive, in the way a doctor's goals are—he wants to heal, renew, nurture, revitalize, repair, revive, mend, soothe, regenerate, and reinvigorate.

This is the main lesson I have learned from Bill Zeedyk over many years of knowing and working with him: a positive solution begins with a positive outlook.

TO LEARN MORE

*Let the Water Do the Work: Induced Meandering,
an Evolving Method for Restoring Incised Channels*
by Bill Zeedyk and Van Clothier.
Chelsea Green Publishing, White River Junction, VT, 2014.

Two lectures by Bill Zeedyk on riparian and wetland restoration can be viewed at: https://www.youtube.com /watch?v=V3d85D4xlbA and https://www.youtube .com/watch?v=cphgauLh32E

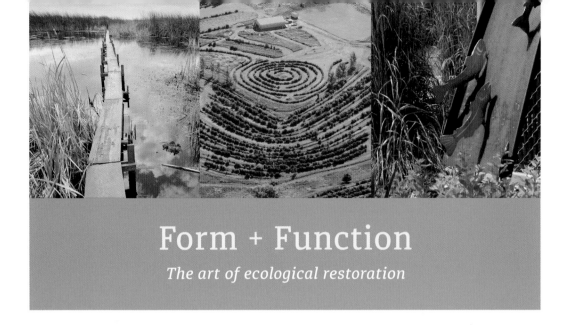

Form + Function
The art of ecological restoration

Harmony, beauty, and aesthetics are all signs of health in nature and in ourselves.

One of my heroes is the conservationist Aldo Leopold, widely honored for his pioneering work in many fields of endeavor, including wilderness protection, wildlife management, environmental education, and even sustainable agriculture. Leopold is best known for his articulation of a land ethic, which is essentially a plea for harmony between land and people, and for the concept of land health, which encompasses the regenerative processes that perpetuate life. But there is another aspect of his deep thinking that has been largely overlooked: beauty is also an important component of conservation and ecological restoration.

This was one of the reasons I took a shine to the creek restoration ideas and methodologies of Bill Zeedyk, a former Forest Service biologist who has become a pioneering restoration specialist. Zeedyk's methods harmonize with the land and its ecological processes—and not coincidentally, the installments he places in creeks to heal them are also attractive to look at. Made of rocks and wooden posts, they have a sculptural feel that verges on the artistic. It is work that integrates form and function on one acre—as Leopold foresaw.

One of Zeedyk's students, Craig Sponholtz, has taken this idea to the next level, transforming stream and upland restoration into an art form. For example, Sponholtz created a log-and-rock structure called a step-down below a wet meadow in a degraded stretch of Grassy Creek, high in the Valle Vidal unit of New Mexico's Carson National Forest, that is both highly functional and very pretty.

Sponholtz arranged zigzagged spruce logs in the creek to make it look like the trees had simply toppled over from the nearby forest. In fact, they were carefully placed by Sponholtz using an excavator. The locally sourced rocks were also fitted around the logs in a way that was pleasing to the eye, and yet I knew that each one had an important role to play in healing the creek. Add in the tufts of sod inserted between the rocks and logs, the pretty rock-lined bowl at my feet, and the burbling sound of cascading water, and you had the recipe for a Zen-like work of art—in a wildflower-strewn valley, nestled between a rock outcropping and a patch of spruce forest, under a canopy of blue sky. What a gallery for a grand opening!

Of course, the step-down structure had a job to do, first and foremost. Its assignment was to save the wet meadow by easing water down a steep stretch of creek without incurring any additional erosion, especially in the event of a big rainstorm. Accomplishing this goal requires knowledge of soils, hydrology, geomorphology, mechanical engineering, and math on the part of the designer, as well as a great deal of field experience (and a soft touch with an excavator), or the structure will fail in its duty. But this is where Leopold came in. What Sponholtz had done on Grassy Creek was take something totally functional and entirely human-constructed and make it look like a natural feature on the land, in this case an attractive log-filled cascade of merry water. It was a wholly practical restoration structure *and* a piece of sculpture. The dictionary defines *sculpt* as "to carve," which is exactly what Sponholtz had done—carved the land into health and beauty.

Craig Sponholtz's log-and-rock step-down in Grassy Creek harmonizes with the land while healing ecological damage. *Photo courtesy of the Quivira Coalition*

Aldo Leopold is an inspiration to Sponholtz as well. His favorite Leopold quote is: "A thing is right when it tends to preserve the integrity, stability, and beauty of the biotic community. It is wrong when it tends otherwise." That's exactly what the log and rock step-down was doing—restoring the integrity, stability, and beauty of the biotic community known as Grassy Creek.

"Beauty forms a large part of our relationship to nature," Sponholtz told me, "and we react negatively when it's absent, as with degraded landscapes. If you're going to spend time and money trying to heal a meadow like this one, which is critically important to the ecosystem, then I think it's best to follow nature's blueprint, which involves an intricate web of interactions that life depends on. Beauty is part of that web, as are water, soil, and plants. You can't have one without the other."

The art of restoration isn't simply about the structure itself, the materials used to make it, or its intended effect on the creek. It's also about what happens before you get started. It's about recognizing opportunities, which requires

This media luna on the Navajo reservation, designed by Sponholtz, spreads out stormwater, reducing erosion. *Photo courtesy of the Quivira Coalition*

careful observation and a clear understanding of nature's regenerative principles. Opportunities can take the form of a boulder or bedrock outcropping, a clump of sturdy vegetation, a low bank, or almost any other feature that can be used to create a unique solution. Once recognized, these opportunities help to create a structure that harmonizes with long-term natural processes. Sponholtz calls these opportunities "sweet spots."

"My goal is to recognize the small opportunities that make a big difference and to act on them," he said. "This is why watershed restoration is endlessly creative and endlessly rewarding."

Of Sponholtz's work, especially lovely is the water-spreading, crescent-shaped structure called a media luna (half moon), which he has perfected into sculpture. Another specialty is an in-stream grade-control structure called a cross vane, which is composed of large rocks carefully arranged in the creek in order to slow down the water's momentum by creating a natural plunge pool.

After taking care to diligently read the landscape of the project site, Sponholtz creates a design that involves as few people and materials and as little dirt moving as possible, while striving for a strong and long-lasting effect. This minimalism is partly about self-expression, but it's also about physical objectives—to heal the creek as simply and effectively as possible. It also makes sense economically, especially to the landowner or agency funding the work. Beauty is woven into the minimalism too, which accounts for the naturalistic feel of his structures.

A storm drain and discharge basin in the city of Santa Fe, designed and built by Sponholtz, blends form and function attractively. *Photo courtesy of the Quivira Coalition*

Sponholtz calls what he does "regenerative earth art." Not only is his goal to heal damaged land for anyone who lives in a watershed (all of us, in other words), he creates structures that become part of the ecological processes that they reignite. By serving as footholds for grass and riparian plants that take over, his structures eventually disappear into the land itself. Best of all, this integration of the ecological and the aesthetic can happen anywhere, even in cities.

"The main misconception that people have about watershed restoration," Sponholtz said, "is that it's something that happens far away in parks and public lands and not something that can be part of everyday life. But everyone lives in a watershed, and I work hard to make the restoration of our home watersheds something that is built into the ways we live and work."

Healing the land is healing ourselves—one acre at a time.

TO LEARN MORE

Visit Craig Sponholtz's website at:
www.watershedartisans.com

For a short lecture by Craig Sponholtz,
see: https://www.youtube.com/watch?v=eCU27aEvEIo

Sweet Spots

Special places where small investments get big returns

I t's human nature to try to solve the biggest crisis, fix the nastiest wound, or confront the most shocking outrage first, when in fact we could get bigger returns if we focused on smaller problems with greater potential—the sweet spots.

A few years ago I visited an example of a sweet spot on a farm on Twitchell Island, in the middle of the great Sacramento-San Joaquin River Delta, east of San Francisco. I didn't travel to the island to see farmland, however. I wanted to see a *carbon* sweet spot in action. On Twitchell, a whole suite of big things had happened on just fourteen acres of wetlands in only a few years and for very little cost.

Thanks to a high density of plant matter and a low rate of decomposition, wetlands are the world's best ecosystems for capturing and storing the carbon from CO_2 in their soils. Their destruction, conversely, releases lots of CO_2 into the atmosphere as these soils dry out and oxidize. Moreover, at least one-third of the world's wetlands are composed of peat, a type of soil created by dead or dying plants that are permanently water-bound. Peatlands, which include bogs and fens, contain 30 percent of global terrestrial carbon but cover only 3 percent of the Earth's land surface, which is a lot of carbon bang for the buck.

The Sacramento-San Joaquin River Delta was once a vast freshwater marsh, thick with tule reeds, cattails, and abundant wildlife. At least six thousand years old, the marsh caught sediment that washed down annually from the Sierra Nevadas, building up soil that eventually extended sixty feet deep in places. When the delta began to be settled in the 1860s, following California's famous Gold Rush, farmers couldn't believe their luck. Because the soil had been often submerged—a consequence of flat terrain, frequent flooding, and tidal action—it had essentially become peat, rich in carbon and other organic minerals.

Crops grew vigorously in the rich soil. Soon, a new gold rush was on to claim land in the delta, drain it, and grow row crops by the bushel load.

Fast-forward to today, and the delta is in big trouble. Innumerable ditches and levees have broken up the marsh into 57 separate islands, 98 percent of which are now below sea level. Pumps work continuously to keep the roots of the crops dry enough to grow and be harvested. Salt intrusion from the bay is creeping inland, threatening not only the crops but the drinking water supply for two-thirds of all Californians and much of the state's agriculture. Not many people know that central California is a vast plumbing project, crisscrossed by a complex network of canals, ditches, and pumping stations. And most of the water in this plumbing system originates in the southern part of the Sacramento-San Joaquin River Delta.

However, the islands are sinking, sea level is rising, and the 1,100 miles of levees that protect it all are feeling the stress, literally. The phenomenon is called subsidence, and it places tremendous hydrostatic pressure on the levees, requiring constant maintenance—and creating perpetual anxiety. What if the levees were breached by a massive flood? What if salt water poured through, ruining crops and drinking supplies?

In 1997, in an attempt to alleviate these worries, a group of scientists led by Robin Miller of the US Geological Service came up with a novel idea: employ nature, not technology, to reverse the subsidence. When the early farmers drained the delta they exposed the peat soil to the atmosphere, causing the organic material that was previously underwater to oxidize rapidly. The carbon in the soil literally blew away, causing the land to compact and subside over time. That's how the islands ended up below sea level—as much as 25 feet in some places. The scientists wondered: Could this process be reversed? In other words, could the land be built back up if the marsh ecology, including periodic flooding, could be resurrected?

A boardwalk across the Twitchell Island study site. Measurements indicated that ten inches of soil had been created in less than seven years in the plots. *Photo by Matthew Grimm/Environmental Defense Fund*

To find out, they implemented an experiment on two 7-acre, side-by-side plots of farmland adjacent to a ditch that bisected Twitchell Island. They flooded the western plot to a depth of 25 centimeters, and the eastern plot to 55 centimeters. Tules were planted in a small portion of each plot. By the end of the first growing season, cattails had colonized both plots (the seeds arriving on the wind), which provided a screen for other plants, including duckweed and mosquito fern. Then things really took off. After just a few short years of annual managed flooding, the western plot had developed a dense canopy of marsh plants, as had the eastern plot, though it maintained some open water.

When the scientists took measurements of the soil after seven years, they were amazed to discover that the soil in both plots had risen *10 inches*—the result of 15 tons of plant material growing and dying per acre per year. This was great news.

"Ten years after flooding," wrote Miller in a peer-reviewed summation, "elevation gains from organic matter accumulation in areas of emergent marsh vegetation ranged from 30 to 60 centimeters [1 to 2 feet], with an annual carbon storage rate approximating 1 kg/m^2, while areas without emergent vegetation cover showed no significant change in elevation."

The researchers next tested the amount of CO_2 that had been sequestered in this new soil as a result of their experiment. They

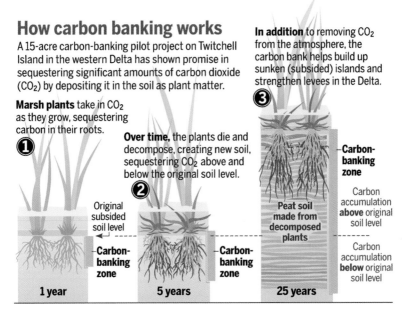

How carbon banking works

A 15-acre carbon-banking pilot project on Twitchell Island in the western Delta has shown promise in sequestering significant amounts of carbon dioxide (CO_2) by depositing it in the soil as plant matter.

In addition to removing CO_2 from the atmosphere, the carbon bank helps build up sunken (subsided) islands and strengthen levees in the Delta.

Marsh plants take in CO_2 as they grow, sequestering carbon in their roots.
❶

Over time, the plants die and decompose, creating new soil, sequestering CO_2 above and below the original soil level.
❷

❸

Original subsided soil level

Peat soil made from decomposed plants

—Carbon-banking zone

Carbon accumulation **above** original soil level

—Carbon-banking zone

—Carbon-banking zone

Carbon accumulation **below** original soil level

1 year   5 years   25 years

How carbon banking works. When a marsh is flooded, drowned plants become a significant source of soil carbon. *Image by Dave Johnson, courtesy of the Bay Area News Group*

suspected that 10 inches of dense, carbon-rich peat soil likely soaked up a lot of atmospheric CO_2—and they were right. In fact, as much as 25 metric tons per acre per year were sequestered in the study plots, according to their analysis. In comparison, a typical passenger vehicle emits 5 metric tons of CO_2 per year. The 14 acres in the study plots sequestered the equivalent emissions of 70 passenger vehicles per year! And that doesn't even count the CO_2 emissions eliminated by not farming the land. And it doesn't count all of the other ecosystem services generated by a functioning marsh, including water purification and wildlife habitat.

The researchers called their project a "carbon-capture farm'—a process also known as carbon banking—and hoped that the project would demonstrate that it is highly feasible to use managed wetlands to sequester carbon and reduce subsidence simultaneously. The key word here is *managed*, which raises another whole set of questions, especially about working at scale. Although the specifics of this project are likely limited to the Sacramento-San Joaquin River Delta, it is nonetheless a very good example of a sweet spot. On just 14 acres, the project demonstrated how to reverse subsidence; reduce the risk of levee failure; sequester a lot of carbon; and provide wildlife habitat, especially for birds on the Pacific flyway. Visiting it taught me that the best way to address the big picture is to begin in places where we can achieve inspiring results quickly and for a low cost.

Sweet spots are all around us, if we know where to look!

TO LEARN MORE

An open-access paper, "Re-Establishing Marshes
in the Sacramento-San Joaquin Delta of California,"
by Robin Miller et al. is available from
Nova Publishers: www.novapublishers.com

For more information about carbon farming
from the US Geological Survey, see: http://ca.water
.usgs.gov/Carbon_Farm/RandD.html

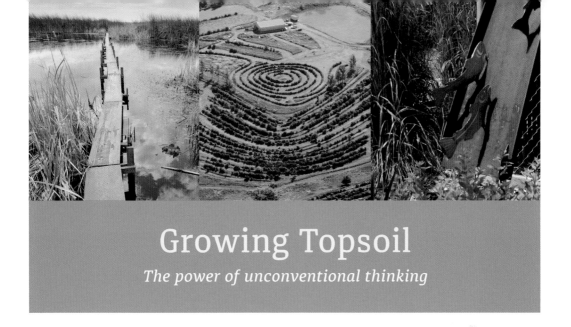

Growing Topsoil

The power of unconventional thinking

I s topsoil a renewable resource or a nonrenewable resource, especially in dry or degraded landscapes?

The answer to this question is important because we're losing topsoil every day—lots of it. Geologically, topsoil is produced by the physical and chemical weathering of rock as plant roots widen cracks made by freezing and thawing action and carbonic acid in raindrops breaks down the pieces into sand, silt, and clay particles (as does the grinding work of glaciers). Just add organic matter—carbon—and voila, topsoil! However, it may take as long as a thousand years to build an inch of biologically active soil through this process, which makes topsoil a nonrenewable resource on human time scales.

Nearly two billion tons of topsoil erode annually from American farms and ranches, primarily due to poor agricultural practices. Most of this soil washes into creeks and lakes and out to sea.

Actually, two billion tons is a big improvement. Twenty-five years ago the amount of topsoil lost annually to erosion was 40 percent higher. The difference is the adoption of a suite of agricultural practices—including the use of cover crops, no-till farming, and regenerative grazing—that reduce the erosive power of rain and wind. The goal of these practices is the conservation of topsoil for the future. In other words, if soil is a nonrenewable resource, the best we can do is slow down its rate of loss.

But what if topsoil was also a *renewable* resource? What if a farmer or rancher could create an inch of biologically active topsoil in a decade? According to conventional thinking, that can't be done—not with chemical-based agriculture, anyway.

Fortunately, unconventional thinkers have had other ideas.

Take Charles Darwin. In his final book, published shortly before his death in 1882, the great scientist focused his research on the lowly earthworm and the role it played in the mystery of soil formation. By conducting a variety of experiments in his backyard over many years, Darwin discovered that topsoil can be expanded (deepened) in only a matter of years, largely as a result of the digestive work of earthworms. This was big news at the time. The idea that soil was biologically alive with critters transforming inert subsoil into rich topsoil by eating and pooping was rather revolutionary. Of course, Darwin had the advantage of living in England, where moist conditions can speed up biological processes. What about drier parts of the world?

Sixty years later, the answer came from another unconventional thinker, this time on the far side of the world and the other end of the celebrity scale.

P. A. Yeomans was a former Fuller Brush salesman in Australia who took a correspondence course in geology and became a mining engineer in charge of large earth-moving projects. In his new job, he carefully studied the way water moved across the land, especially gravity flow. A restless experimenter, Yeomans decided after World War II to trade mining for agriculture and purchased a farm in New South Wales, where he began to test his unconventional ideas of water and land management, including the "keypoint" concept. A keypoint is the precise spot in a small valley or drainage where water slows down enough to be directed underground via a narrow "Keyline" ditch dug on the point's contour line. His objective was to get as much water into the soil as possible, thus recharging the plant life, especially if the soil was degraded or compacted.

It wasn't just intellectual curiosity at work, however. When a prolonged drought hit Australia, contributing to a devastating fire on Yeomans's farm that killed his brother-in-law, he vowed to drought-proof his property—and by extension, all of Australia! Explaining his goals, Yeomans wrote, "The landman's job is not so much to conserve soil as it is to develop soil and to make it more fertile than it ever was."

Yeomans pioneered two paths toward his goal. The first involved a tool. On a visit to Texas, he watched a chisel plow in action and realized that with modifications this plow was ideal for "ripping" key-line contours across farms and ranches. A chisel plow cuts a narrow, deep furrow (8 to 12 inches) without turning over the dirt and is used primarily to loosen rocky or compacted soils. Yeomans recognized its potential for encouraging water and oxygen infiltration in the soil—keys to "revving up" biological life underground. Healthy soil is chock full of microorganisms (trillions of them) and like all forms of life they need water, oxygen, and food (carbon) to thrive. But if soil becomes compacted, all life underground suffers. To alleviate these conditions,

Whirlwind Farm, in southwestern New Mexico, before Keyline treatment. Whirlwind Farm after Keyline treatment by a Yeomans Plow and summer rain. *Photos courtesy of Owen Hablutzel*

Yeomans designed, tested, and patented what is today known as a Yeomans Plow for exactly this purpose.

His second innovation was conceptual, what scientists today call "resilience thinking"—how to bounce back ecologically or economically from a surprise or shock. Yeomans developed a whole-systems approach to his farm, insisting that close attention be paid to all parts of the land under management, including proper grazing by livestock. Goal setting, design, testing, and retesting needed to be incorporated into every farming enterprise, he said, and appropriate scales needed to be respected. The primary ecological objective of all this planning was to increase the regenerative capacity of the land, and to do that people needed be treated as an integral part of any management system. Although Yeomans probably didn't use the word *resilience* to describe his goals, it certainly describes his intentions.

These concepts, by the way, are the foundations of the nature-based design process called permaculture, developed by fellow Australians Dave Holmgren and Bill Mollison in the 1970s.

So, do Yeoman's innovations actually build topsoil?

Yes, says Owen Hablutzel, an expert in whole-systems farming and ranching. The Yeomans Plow is a good tool for fixing a damaged water cycle, Hablutzel told me, by preparing compacted soil for rain. One or two 8-inch-deep rips by the plow below the labile (top) layer of soil jumps up the level of biological activity. The chances are good, he said, that the plow can increase soil carbon as a result.

"Among farm and ranch clients, and Keyline projects I've known personally in New Mexico, Texas, Nevada, and California," Hablutzel said, "we have seen increased topsoil and soil organic matter, restored pasture on abused former cropland, reduced soil loss from erosion,

A Yeomans Plow "rips" a 10-inch-deep furrow in compacted soil, allowing water to feed soil microbes, thus building topsoil. *Photo courtesy of Owen Hablutzel*

reduced or eliminated invasive brush, longer growing seasons, and greatly increased soil moisture, soil life, and overall fertility."

The Keyline strategy works well in dry country too—perhaps *especially* in arid lands. That's because every drop of water is precious, particularly in a drought, and any method that can get more of the wet stuff to the roots of plants, the better.

Despite these successes, however, Yeomans's ideas remain unconventional for many in agriculture. Partly it's an image problem (he wasn't considered a "real" farmer by his peers) and partly it's a lack of scientific scrutiny to back up claims at this point, a situation that will hopefully change in the near future.

Nevertheless, in this era of rising environmental and social stress, we need more unconventional thinking—and quickly!

TO LEARN MORE

For P. A. Yeomans's books and more information
on Keyline, see: www.keyline.com.au

View a lecture by Owen Hablutzel on
resilience science and Keyline design at:
www.youtube.com/watch?v=V3twLVn7nss

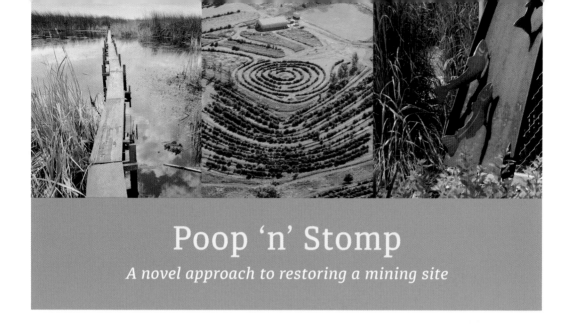

Poop 'n' Stomp
A novel approach to restoring a mining site

This is a story about how life was created from lifeless soil.

Pulling up at a stoplight in Globe, Arizona, one day, I casually glanced through the windshield and saw 30 cattle grazing peacefully together on the slope of a tall hill. There was nothing unusual about the image—except that the hill was actually a huge pile of mine tailings, where the waste rock from decades of open-pit copper mining nearby had been hauled and dumped. From a distance, the pile looked like a giant steep-sided ziggurat—an ancient Mesopotamian edifice that rises in levels from a massive base—only with cattle grazing on its side! I imagined that this sight might seem incongruous to many people, but I knew what was going on.

It was a "poop 'n' stomp."

Grass grows on mine tailings in Globe, Arizona, thanks to the poop 'n' stomp effect of cattle grazing (the cattle are visible on the far left). *Photo by Courtney White*

A century of unregulated, poorly designed, and poorly executed mining has caused a litany of well-documented environmental damage around the world, and one of its biggest challenges is the waste left behind. Since the rock is excavated far below the surface, it is essentially sterile—colorful, perhaps, but lifeless. This lack of organic matter means the tailings, when piled high, quickly erode (especially after a torrential summer thunderstorm)—causing all sorts of mayhem downstream. Accumulating sediment, for example, can create maintenance headaches for reservoirs. As a consequence, regulatory agencies often engage in arm-wrestling contests with mining companies to slow or stop the erosion caused by their operations.

That wasn't a problem with the hill I observed, however, because it was covered with vegetation. For confirmation, I drove to the eastern side of the ziggurat, where, as I expected, I saw grass—lots of it. I knew why. The cows had worked over the tailings recently and apparently it had rained in the interim. I knew this because we had tried something similar years ago on mine tailings in New Mexico, albeit on a much smaller scale. Our goal had been to grow grass—life—on largely lifeless soil, principally using the tool of grazing animals. It worked too, as I'll explain.

In early 1999, I received a phone call from an Environmental Protection Agency administrator in Dallas, Texas, who said they had some extra money in a Clean Water Act account and asked if I might be interested in conducting a restoration project with it. He knew that our little nonprofit, which focused on the ecological benefits of good livestock management, was eager to implement demonstration projects in the region. When he suggested a mining reclamation project, I said, "You bet!" That's because I knew who to call.

I had recently met Terry Wheeler, a feisty and outspoken rancher from the Globe area who had pioneered a mine-reclamation strategy that used only livestock, hay, grass seed, electric fencing, a portable water source, and a human worker. His idea was as simple as it was brilliant: build a small paddock with electric fencing on a patch of eroded slope, spread grass seed across the ground followed by hay, turn the cows into the paddock for a few days, and watch as they pressed the seed and bits of hay into the ground with their hooves while eating. Add the bodily functions of the livestock, rain, and time, and presto! Green grass.

It was no different, Terry liked to observe, than the instructions on the back of a packet of seeds that you buy to plant in your garden: *press seed firmly into soil.* Just add water. The only additional variables in this case were the hay (a carbon source), the nature of the fertilizing process, and the seven-hundred-pound animals who did most of the work, even on a forty-degree sterile slope. The employment of portable electric fencing and a moveable water source (usually on a truck) meant that Terry and his cattle could work their way across the face of the wastepile in a methodical manner.

Cattle stomping on mine tailings near Cuba, New Mexico, recalling the instructions on a packet of seeds for a garden: *press seeds firmly into soil.* *Photo by Courtney White*

What Terry was doing was such a novel idea at the time that no name for it existed, so I made one up: poop 'n' stomp.

When Terry first developed the idea, he approached the owners of a huge copper mine near his home. They were both curious and skeptical, he told me. Many traditional mine-reclamation strategies involve costly combinations of water pipelines, mechanical sprayers, chemical fertilizers, diesel-powered machines, and human labor. The goal is to stabilize the tailings so they won't erode into a nearby creek, and if the process is not designed properly, implemented correctly, and maintained adequately, then all that work and money is often literally washed away in a few years. So when Terry told the mine owners that he could reclaim one of their massive tailings for less money and with better results, using an organic process to boot, he got their attention. Their skepticism kicked in when he said he would do the work with cattle.

"One mining executive," Terry told me, "liked to joke that they should line up BBQ grills at the bottom of the slope for all the cattle that would come tumbling down."

The cattle didn't come tumbling down, of course. They did just fine, pooping and stomping their way back and forth across the pile under Terry's guidance, pressing the grass seeds firmly into the ground

When rain repeatedly hits bare ground, it can effectively seal the top layer, creating what's called "capped" soil. *Photo by Courtney White*

with their hooves. When the rain came and the grass grew, the jokes stopped, Terry told me.

Terry called his cattle *FLOSBies*—four-legged organic soil builders.

Using the EPA funding, I hired Terry to manage a similar project across a 20-acre patch of eroding soil on an abandoned copper mine near Cuba, New Mexico, with the permission of the private landowner. Over the course of two summers, Terry's small herd of *FLOSBies* poop 'n' stomped those 20 acres back to life. Winter snows and spring rains caused the steep slopes to grow a great deal of grass. Soil stabilized, gullies healed, rain soaked in instead of running off, and the ground turned green during the summer. Various agencies, including the EPA, were pleased.

However, I wasn't just thinking about mine waste with this demonstration project. I was intrigued by the possibility of using cattle in the service of environmental restoration generally. One summer, for example, a forest fire near Santa Fe burned so intensely that it essentially sealed shut the forest soil. Water couldn't penetrate the seal, which meant that grass seed sprayed aerially by the US Forest Service over the burned area, as a way to jump-start the recovery process, washed away with the first thunderstorm. Terry's idea was to use a herd of cattle to break the sealed soil with their hooves, allowing the seed to reach fertile ground. I actually called the Forest Service to see if they would let us try a small demo project on the burn, but the answer, alas, was a polite "no thanks."

It's not just cattle—*FLOSBies* can come in all four-legged shapes and sizes, even wild ones. All you need to get started is an eroded sweet spot where a small investment will yield big returns and the desire to make it healthy again.

For a society fixated on technical and petroleum-based solutions to its multiple problems, it was inspirational to discover an organic alternative that could be effective and regenerative.

TO LEARN MORE

For information on land restoration using cattle,
see: Holistic Management International,
http://holisticmanagement.org, and
The Savory Institute, http://www.savoryinstitute.com

Here is a related story from
High Country News: http://www.hcn.org/blogs/goat
/cow-stomp-using-cattle-to-reclaim-mine-land

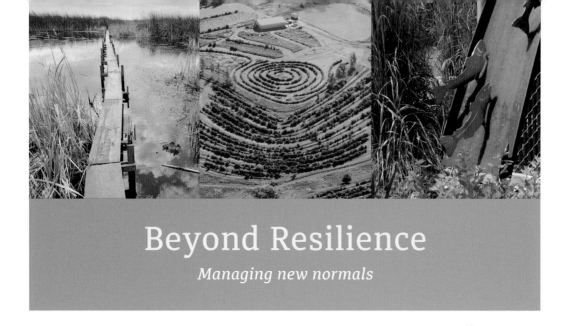

Beyond Resilience
Managing new normals

Restoring land to health means trying to return it to something like normal ecological conditions. But what if the definition of *normal* changes in the meantime?

An ecosystem's capacity to absorb a shock, such as a drought, flood, or forest fire, and then bounce back as quickly as possible is called *resilience*. Since it's a critical part of ecosystem health, ecologists have made a big effort to understand what constitutes "normal" conditions in order to help a system be as resilient as possible, especially if the shock has been caused by humans, such as overgrazing by cattle. But what if a system's definition of normal changes? What if a region's annual precipitation dropped by half—and stayed there? Or when the rains did fall, they came as unusually large flood events or at the wrong time of year? What does *resilience* mean in this context?

It's not an abstract question. Under climate change, scientists tell us, we'll be experiencing all manner of new normals. For restoration purposes, this means we need to search the management toolbox for practices that go beyond short-term resilience and allow an ecosystem to endure long-term deviations from normal conditions.

What would those practices be? Mike Reardon has an idea.

Since the late 1990s, Reardon has used a wide variety of land restoration tools on his family's 6,500-acre Cañon Bonita Ranch, located in northeastern New Mexico. These tools include tree removal, brush clearing, prescribed fire, planned grazing, erosion control, riparian restoration, water harvesting, dam building, and ranch road repair—all in service of restoring ecological health to the land after decades of mismanagement by previous landowners. Reardon's overall goal is to support a multitude of diverse wildlife on the property and his work has been highly effective in this regard. Today, however, he faces a new

Abundant and diverse grasses return to the Cañon Bonita Ranch as a result of a combination of land restoration and management strategies. *Photo by Tamara Gadzia*

challenge: How do you maintain forward progress when prolonged drought limits the use of certain tools?

In 1997, an expert with the USDA's Natural Resources Conservation Service told Reardon that there were "too many trees" on his ranch. This was news to Reardon, who lives in Albuquerque and readily admits to being a novice about land health when he began managing the ranch. Too many piñon and juniper trees, the expert said, meant a reduced amount of open, grassy habitat for wildlife. In the past, nature corrected this situation with periodic, lightning-sparked wildfires that would thin out the trees, allowing the land to bounce back with perennial grasses. However, a century of fire suppression by landowners and cooperating agencies across the region, coupled with poor livestock management, eventually eliminated the land's grass cover, resulting in widespread tree encroachment.

To reverse this situation, Reardon focused first on reducing the density of piñon and juniper trees on the property. His original tools were handheld loppers and a chainsaw. Then came a spin trimmer, a front-end loader, and a Bobcat skid-steer. Next, Reardon hired a professional woodcutting crew from Mexico. To date, nearly three thousand acres have been cleared on the ranch, though some stands of trees were left for wildlife.

Next, during the years when grass (and rain) was abundant, Reardon alternated the use of two other tools to further reinvigorate the grasslands: prescribed fire and planned grazing. With the assistance of neighbors and fire experts, Reardon has completed two controlled burns, ten years apart, which effectively suppressed tree seedlings. Reardon also employed the tool of high-density, short-duration grazing by cattle during the vegetative dormant season (December through March). This "living fire" recycles old grass into cattle manure, which helps to build grass cover.

All three tools worked. Grass came back with a flourish, teaching Reardon an important lesson.

"I learned that bare ground was enemy number one," Reardon said, "so I do everything I can to get grass to grow. And not just any grass, I want perennials and I want as much diversity as possible."

The next job for the resilience toolbox was water. In order to create more surface water for wildlife to drink, as well as grow a year-round supply of nutritious food, twelve earthen dams and four metal tanks (with windmills) were repaired, modified, or constructed across the ranch. He also implemented a five-phase wetland and riparian restoration project that employed many of the innovative practices pioneered by specialists Bill Zeedyk and Craig Sponholtz.

They designed and implemented treatments for a two-mile stretch of Cañon Bonito Creek, which ran through the center of the ranch. Their goals were to decrease stream bank erosion and downcutting and to raise the water table. They also wanted to reconnect the creek to its floodplain in order to re-wet adjoining wet meadows and increase the amount of live water. They also hoped to increase forage species, including wetland vegetation, and increase cover for wildlife. There was even a plan to harvest water from ranch roads using a variety of techniques, including redesigned road crossings and water-harvesting rock structures in canyon side channels.

Reardon also implemented a detailed monitoring program on the ranch in order to see how changes were progressing. This included vegetation and bare-ground monitoring, moisture data collection, wildlife population surveys, and photographic documentation, including sixty photo points along Cañon Bonito Creek alone.

The message of the monitoring data was clear: conditions were improving. Under Reardon's management, the ranch progressed from a monoculture of blue grama grass to hosting a diversity of more than 55 different grass species. Dry springs began to flow again and wildlife populations shot up by a factor of ten. Despite a drying trend that began in 2002, deer, elk, and wild turkey populations continued to rise and things seemed to be returning to normal. It looked like Reardon had succeeded in rebuilding resilience on the ranch.

Except—the definition of *normal* was changing. The drought, for example, went on and on—and still goes on.

Today, year-round water in the Cañon Bonito Creek is rare, though there is still a steady trickle in the spring area. A relict population of ponderosa pines is dying, along with piñon and juniper trees. Small populations of perennial grasses, previously restored, are now dying as well. And wildlife populations are in decline—wild turkey populations have dropped by 75 percent. As for the land management toolbox, persistent drought means that prescribed fire is off the table

The new normal of big flood events on the ranch. What is the best strategy for coping with these new conditions? *Photo by Tamara Gadzia*

and grazing by cattle is limited to selected areas of the ranch.

Reardon has learned the hard way that getting "beyond resilience" is easier said than done.

On the good news front, there is still plenty of ground cover holding the soil in place, capturing "airmail topsoil," as Reardon puts it, during local dust storms, as well as any raindrop that falls from the sky. The wetland and riparian restoration work have kept the ground moist where otherwise it might have gone dry. It also helps to dissipate the destructive forces of unusually big flood events, such as one the ranch endured in September 2013, when nearly five inches fell in a matter of hours. Thanks to all the vegetation that had grown along the stream banks, the effects of that flood were not nearly as devastating as they would have been otherwise.

For Reardon, the whole experience points to important lessons learned for the new normals of hotter, drier conditions and chaotic moisture events.

"Use your time effectively," he said, "focus on sweet spots, have a plan, pull together a diverse group of supporters and professionals, be willing to listen and learn, trust the data, be willing to admit mistakes, be proactive, become land literate, and get ready for the next storm—dust, rain, snow, whatever Mother Nature brings. It will rain again!"

Sage words as we move deeper into the twenty-first century!

TO LEARN MORE

Resilience Thinking: Sustaining Ecosystems and People in a Changing World by Brian Walker and David Salt. Island Press, Washington, DC, 2006.

"Unprecedented 21st Century Drought Risk in the American Southwest and Central Plains" by Benjamin I. Cook, Toby R. Ault, and Jason E. Smerdon. An online article in *Science Advances,* 1, no. 1. (2015): http://advances.sciencemag.org

PART FIVE
Wildness

The Bee's Knees

Rebuilding from the bottom of the food chain up

When it comes to creating a regenerative agriculture and combating climate change, we have allies in nature, particularly certain keystone wildlife species whose restoration and protection will enrich all our lives.

A keystone species is one that has an outsized role in the overall health of an ecosystem, like the keystone in an arch of a doorway. Pull it out and the door—and possibly the wall—collapses. One such keystone species is the gray wolf, which was nearly exterminated in the twentieth century. The reestablishment of the wolf to its former range in the northern Rockies beginning in the mid-1990s is one of the great success stories of modern conservation. It's also a prime illustration of "top-down" biology at work. Wolves are an apex carnivore species whose presence creates a cascading effect downward through the food chain. This keystone effect is an important one for ecosystem health—though you might feel otherwise if you were the slowest member of an elk herd!

But what about species in trouble at the bottom end of the food chain? Some, such as monarch butterflies, get lots of attention and are easy to love, but what about bees and other insect species that have a keystone role to play in ecosystem health—as well as human food production—but are much more "charismatically challenged"? Wouldn't restoring their populations and habitat have a positive cascading effect *upward*? More to the point, could a bottom-up approach be complementary to top-down conservation work that might otherwise not function as well on its own for lack of abundant prey? If so, is it worth the time and money to educate the public to support pollinator habitat recovery through "Save the Bee" campaigns?

Early research suggests that the answer to the first question is yes—building the base of food chains has a significant effect all the way up the line, including with ongoing predator recovery efforts. The answer to the second is "Hell yes!" because healthy bee and butterfly habitat attracts other insects that provide food for game birds and omnivorous mammals, including us. Bumblebees, butterflies, and nectar-feeding bats and hummingbirds, among others, are inextricably connected to resilient and productive food chains, healthy water and mineral cycles, and even to the creation of soil carbon via herbivore grazing and sustainable farming.

Of course, as with the wolf, native insects have every right to exist and be healthy for their own sake. Linking them to human food production just increases their odds of survival.

To test these questions, a pioneering effort at bottom-up restoration has begun in the Sonoita Creek watershed of southern Arizona, home to over 300 species of native bees, 180 different butterflies and moths, 14 species of hummingbirds, and 2 types of nectar-feeding bats. This collaborative effort is called Borderlands Restoration, and its aim is to restore and protect ecological spaces for pollinators and their nectar sources. Food chain restoration, a term coined by ecologist and local resident Ron Pulliam, starts by repairing damaged creeks and streams in the area, crucial habitat for pollinators. After that, nectar- and fruit-bearing plants are planted along the healed watercourses, which in turn supports pollinators, frugivores (fruit eaters), herbivores, and predators.

The next step in the restoration process is to design and plant linear corridors of fast-growing, closely spaced perennial bushes or shrubs called hedgerows along the edges of fields, orchards, and pastures. As in, *lots* of hedgerows. Bees love hedgerows, and so do ladybugs and butterflies. Research shows that beneficial insects are most active within 120 feet of such perennial habitat. Farmers have known this for a long time, which is why the cultivation of hedgerows is an ancient human activity. Alas, with the advent of industrial agricultural practices, including the use of tractors, chemical pesticides, and fence-to-fence food production, hedgerows have taken a severe beating around the world (not to mention the soil beneath them). And with their increased scarcity came the steep decline of native pollinators.

Recent news headlines about pollinator declines have focused on the domesticated honeybee, populations of which have dropped by at least 33 percent since 2007, triggered by deadly viruses, aggressive mites, and widespread use of a neuro-based toxic chemical class called neonicotinoids (which are banned in Europe). But Gary Nabhan, a

University of Arizona agro-ecologist and coauthor of *The Forgotten Pollinators*, notes that declines have been just as precipitous for a variety of native pollinators in North America as well, including five species of bumblebees and, to a lesser extent, hummingbirds and bats.

These declines are significant because the pollination services provided by honeybees to American agriculture are estimated to be worth at least $30 billion a year, according to the USDA. That's what it would cost if honeybee hives had to be imported to every farm in the nation (one honeybee hive can cost a farmer $150 to $200). In California, 1.5 million hives are required annually to pollinate the state's vast fruit and vegetable crops—if hives can be found. In Europe, a new study shows that the region is 13 million bee colonies short of what's needed to pollinate its crops.

Insects of many kinds love hedgerows, including native bees, ladybugs, and butterflies. Placing a hedgerow close to a farm field increases rates of pollination. *Photo by Stephen Lavery/Shutterstock*

Which brings up a second concern: subsistence. As Nabhan and his coauthor Stephen Buchmann have calculated, one in every three bites of food consumed in the United States depends directly on insects for pollination, including apples, apricots, avocados, asparagus, broccoli, blueberries, carrots, celery, cherries, cucumbers, citrus, pumpkins, squash, watermelons, olives, pears, peaches, onions, raspberries, and sunflower seeds, and all sorts of nuts. Without pollinators, the fruit and produce sections of our grocery stores would look very different indeed!

The Borderlands Restoration project aims to reverse the downward trend among pollinators in its area by healing creeks and coordinating the planting of diverse varieties of native shrubs and flowering plants, each with a different but complementary ecological purpose. This includes native vines and perennial native milkweeds. Other pollinator-friendly strategies include fences made of dried plant stalks, rainwater-harvesting structures, water-efficient irrigation practices, and anything else that extends the flowering season and keeps beneficial insects around as long as possible.

It's all good stuff for the economy as well.

"Our vision is that the return of formerly forgotten pollinators will not only curb the ongoing extinction of ecological relationships that plagues

One in every three bites of food consumed in the United States depends directly on insects for pollination, including olives, apples, apricots, and avocados. *Photo by Olgysha/Shutterstock*

the continent today," wrote Nabhan, "but will also return economic health . . . [to] the now-impoverished borderlands region, one in which new jobs restoring productive habitat on farms, in native plant nurseries and at nature-tourism destinations would be most welcome."

In particular, he wants to demonstrate that biodiversity conservation on working landscapes of the Southwest can alleviate the crippling human poverty and food-security levels of border counties, which are nearly twice the national average. The key, says Nabhan, is linking research, collaborative conservation, and sustainable food production together so that each can learn from the other.

There are challenges to overcome in implementing food chain restoration, of course, including hungry deer, persistent drought, and tenacious weeds. Getting the right wildflower mix in a particular place can also be difficult. Native seeds are often expensive, and newly planted hedgerows can take as long as eight years before becoming useful to pollinators. Still, the benefits outweigh the costs, as habitat restoration projects and collaborative conservation efforts such as Borderland Restoration demonstrate. And like the reintroduction of the wolf, the effects of this good work are cascading.

In this case, from the bottom up.

TO LEARN MORE

For more information about bee-friendly farming and conservation nationwide, see the website of the Xerces Society: www.xerces.org

The Forgotten Pollinators by Stephen Buchmann and Gary Nabhan. Island Press, Washington, DC, 1996.

Meet the Beetles

The return of nature's sanitation crew

One of nature's most important and overlooked carbon farmers is also an ancient symbol of regeneration and renewal: the scarab.

It's a beetle, a member of the family *Scarabaeidae*, which includes more than 30,000 different species, part of the order *Coleoptera*, which encompasses 400,000 species of beetles (out of the 4 to 8 million still to be classified), constituting roughly 25 percent of all known animal species on the planet. That's a lot of beetles! Too many to keep in mind, so you're forgiven if you hadn't given them much thought. There's one type, however, that definitely deserves our attention: the dung beetle.

It certainly caught the attention of the ancient Egyptians, who elevated the lowly dung beetle to the status of a god— and for good reason. Dung beetles united three sacred elements of their culture: sun, soil, and cattle. Scarabs fly to the dung patties created by cattle and disassemble them within hours, usually by rolling the manure into brood balls— where the beetles lay their offspring—and then burying the balls below ground in

The ancient Egyptians elevated the dung beetle to the status of a god because it united three sacred elements of their culture: sun, soil, and cattle.
Photo by King Tut/Shutterstock

179

tunnels and chambers where the nutrients nourish soil microbes. The ancient Egyptians knew this activity was critical to maintaining the health and fertility of the soil on which their civilization depended, which may be why they revered the dung beetle on a level with Osiris, the god of the underworld.

Alas, the scarab is not so revered today. In fact, dung beetle populations were nearly hammered into oblivion in the mid-twentieth century by the pesticides and insecticides of industrial agriculture. Only in recent years has their benefit to nature and agricultural ecosystems been rediscovered, including the role they can play sequestering atmospheric carbon in soil. It's also been estimated that dung beetles can save farmers billions of dollars every year. How?

The story starts with a fly—the horn fly in particular.

Most people don't realize that manure (dung) is a coveted resource in nature, fought over by many creatures, including the pests and parasites that literally "bug" cattle and other livestock. This includes the horn fly (*Haematobia irritans*—or blood-loving irritant) which arrived on American shores from Europe in 1887. The flies lay their eggs in cow pats and the larvae are incubated there (for as little as five days) until they transform themselves into new adult flies and emerge to begin their torment. Among other maladies, their persistent biting can cause infections in cattle. A century ago, however, horn flies were not the scourge they became for a simple reason: dung beetles eliminated the manure before the eggs could hatch. A bevy of beetles can bury a field of fresh manure patties in a matter of hours—no dung, no flies!

This natural balance changed dramatically after World War II when farms, rangelands, and animals began to be sprayed with various synthetic compounds in the name of pest and parasite "control." Not coincidently, dung beetle populations dropped dramatically (being a "pest" after all), leaving a lot of poop sitting on the ground. Horn fly populations exploded.

Flies can also serve as vectors for a variety of serious diseases that infect humans, including typhoid, cholera, amoebic dysentery, and tuberculosis. One cow patty can house as many as 450 different insect species and one pair of flies can parent as many as 1.5 million new pairs in as little as 14 weeks. Flies can quickly develop resistance to insecticides as well. For all of these reasons, in the early 1970s a handful of researchers and cattle ranchers decided to reject the application of ever-more chemicals and opted to bring back the sacred scarab instead.

Lead by US government entomologist Truman Fincher, an energetic effort began to establish viable populations of two species of dung beetle, one imported from Europe (*Onthophagus taurus*) and one from

Dung beetles can bury a large amount of manure in a matter of hours, carrying carbon underground and reducing infectious diseases caused by flies. *Photo by Stacey Ann Alberts/Shutterstock*

Africa (*Onthophagus gazella*), the latter via Australia where livestock producers were experiencing similar problems. In Africa, research had shown that an elephant dung pile supported 48,000 beetles, who buried the dung underground within hours. One beneficiary of this work was Texas rancher Walt Davis, an early pioneer of high-density, short-duration cattle grazing, which he found to be ideal for the cultivation of dung beetles that trailed his herd of cattle like camp followers trailing an army of soldiers. When he quit using chemicals on his ranch in 1974, the scarab moved in.

"Those beetles really got to work," Davis said in an interview in *Dung Beetles and a Cowman's Profits* by Charles Walters. "In a paddock just vacated by a herd . . . in 48 hours there was no manure. It was gone!"

It was another example of returning to nature's way of doing things, in this case dung removal.

According to Fincher, few people realized the significance of the dung beetles to ecosystems. Beetles are nature's sanitation crew, he insisted. Their quick burial of dung hastens its decomposition, prevents the loss of nutrients, aerates the soil, and increases the depth of soil containing organic material. That sounds like a recipe for building soil and sequestering carbon.

Not only do dung beetles transport carbon, nitrogen, and phosphorus underground when they remove manure, feeding the microbes a rich diet of organic food, their tunnels increase porosity in the soil, which means more water and oxygen reach the microbes as well, revving up their tiny engines. This increases storage of carbon in the soil, with important positive implications for watershed health, plant growth, food production, pollution abatement, and climate change. And all done for free—by nature!

In his book, Charles Walters points out that *Onthophagus gazella* was released precisely as the natural food and organic agriculture movements began to pick up steam in the US, reflecting a desire for nontoxic approaches to food production that continues to this day. "The mere existence of dung beetles," wrote Walters, "is a greater guardian of the organic red-meat supply than all the inspection certificates and agencies of verification can account for."

Then there's the comic sight of beetles flying to fresh dung as if directed by radar. "No one can say that dung beetles are good flyers," wrote Walters. "When their encased wings are uncovered like some secret weapon in a military silo, they rise up almost helicopter style, then lumber along like an early Wright Brothers plane."

Alas, industrial agriculture and its allies were not so amused. The news that their chemicals were killing critters deemed essential to the health of rangelands was not welcome. Infamously, Truman Fincher was forced into early retirement by the US government at the behest of Industry, according to Fincher himself. His research was put on hold and his laboratory samples destroyed. The lowly dung beetle has struggled to regain its proper place in the ecosystem ever since. Fortunately, it's making a comeback, thanks in part to rising interest in regenerative agriculture.

Hopefully, one day the scarab will return to its former lofty status!

TO LEARN MORE

Dung Beetles and a Cowman's Profits
by Charles Walters. Acres USA, Austin, TX, 2008.

For an informative TED talk on the dance of
the dung beetle, see: https://www.ted.com/talks
/marcus_byrne_the_dance_of_the_dung_beetle

Beavers as Carbon Engineers
Another reason to admire this hardworking rodent

Of all the good things beavers do, the least appreciated may be their role as wetland carbon engineers.

Biologists have long considered beaver to be a keystone species, estimating that 85 percent of all wildlife in the American West at some point in their lives rely on the ponds and riparian habitat that beavers create. For example, beaver ponds are important nurseries for fish, including many rare and endangered species. And it's not just wildlife that benefit from our industrious friends. According to the EPA, beaver ponds allow wetland microorganisms to detoxify pesticides and other pollutants, producing cleaner drinking water for people and reducing the cost of water treatments downstream.

The beaver is the largest rodent in North America. It weighs 40 to 50 pounds and has a scaly, paddle-shaped tail and four buckteeth, two on top and two on the bottom. These incisors never stop growing, which means beavers need to keep them filed down by gnawing on trees and other woody objects. Beavers have webbed feet, dexterous hands, and transparent lids that cover their eyes when they swim. They also have a slick coat of fur and guard hair that enables them to live in a wide variety of ecosystems—a quality, unfortunately, that also made them valuable for high-quality pelts, including desirable hats for fashionable Europeans.

In nature, beaver ponds are oases of life. Not only do they provide drinking water for wildlife, but their still waters also harbor a wide variety of aquatic species. Their edges can be especially rich in plant life, including brightly colored wildflowers. Wetlands created by beaver dams are among the most biodiverse ecosystems in the nation, providing essential habitats for plants and animals that would otherwise struggle to survive. As oases, they also provide aesthetic and

Beavers are carbon engineers because a dam can trap as much as one meter of sediment per year and be occupied for as long as 50 years. *Photo by A. J. Gallant/Shutterstock*

spiritual values to people, especially when it means getting a chance to spot an elusive beaver swimming to its den.

Of all the good things beavers do, however, the least appreciated may be their role as *carbon engineers*. By one estimate, as much as one meter of sediment per year is caught behind beaver dams or stored in meadows, and some dams can be occupied for as long as 50 years. Many are large as well, often stretching 1,500 feet long. In 2010, researchers in northern Alberta, Canada, discovered the world's biggest beaver dam, which at nearly 2,800 feet is twice the length of Hoover Dam! Beaver dams also create wetlands around their edges, many of which become carbon-rich meadows over time. In the Upper Mississippi-Missouri River Basin, researchers say, there were once more than 50 million acres of beaver ponds. The total today is down to roughly 500,000 acres, but that's still a lot of carbon sequestration going on, with the potential for much more.

Unfortunately, when beavers are killed or trapped for removal and their dams fall apart, a cascading series of unhappy changes occur, including decreased riparian stability, lowered water tables, higher and more frequent flooding, reduced wetland acreage, degraded habitat for wildlife, diminished water quality, and less resilience to the effects of drought—not to mention all the carbon that is released back into the atmosphere when carbon-rich soils are exposed by erosion.

And we've trapped a lot of beavers over the years.

Before the arrival of Christopher Columbus, it's estimated that 300 to 400 million beavers existed in North America, or roughly 10 to 50 beavers per mile of stream. Today, only 6 to 12 million beavers remain in their original habitat. The decrease wasn't only because of the demand for pelts and hats. In the 1820s, the British-owned Hudson Bay Company sent trappers fanning out across the Pacific

Northwest with orders to kill every beaver family they could in order to discourage American territorial ambitions. No beaver, went the company's logic, meant no "inducement to come hither," as one official put it. The result was the near extermination of beavers from an area the size of France.

The entire beaver trapping/killing episode was a tragedy of epic proportions. Scientists directly link the removal of so many beavers across the American West to the widespread degradation of watersheds that we see today, which is why some consider their near annihilation to be the region's greatest environmental disaster.

Beavers' tree-cutting, dam-building ways haven't endeared them to landowners, however, especially ones who fail to see the ecological benefits of their busy work. Fortunately, this "varmint" attitude among rural residents has been changing somewhat in recent years, as landowners begin to understand that beaver dams keep water on their land longer. Ponded water, for example, will infiltrate the banks of creeks—sometimes called wicking—where it is stored until water levels drop, usually due to drought. When that happens, the water is released slowly over time back into the creek.

What about the nuisances that beavers create in road culverts? A beaver dam beneath a road can cause serious problems, as any landowner can tell you. The answer is easy: install a Beaver Deceiver. It's a carefully constructed fence that discourages beavers from building a dam. Beavers are stimulated by the sound of running water, which inspires them to start gnawing on nearby trees. The farther away a beaver can be kept from these stimuli by the fence, the more likely it is to be "deceived" into leaving the culvert alone.

There's another reason to put these wetland carbon engineers back to work: adaptation.

From prehistoric times to the present, human societies have successfully adapted to the challenges of a changing region, including periods of drought. However, we are entering an era of unprecedented change brought on by new climate realities that will test our capacity for adaptation as well as challenge the resilience of the region's native flora and fauna.

Beaver are a keystone species. Their removal can precipitate a cascading sequence of ecological degradation in watersheds. *Photo by Nancy Bauer/ Shutterstock*

Beavers can help. Here's a list of a beaver dam's resilient attributes, borrowed from the Seventh Generation Institute, a nonprofit that works to restore beavers to their rightful role. A beaver dam:

- Slows snowmelt runoff, which extends summertime stream flows and restores perennial flows to some streams;
- Slows flood events, which could otherwise incise stream channels;
- Contributes to the establishment of deep-rooted sedges, rushes, and native hydric grasses, which buffer banks against erosion during high flows and provide shade to creeks and streams, reducing water temperature;
- Elevates the water table, which can subirrigate nearby land (including farmland);
- Increases the amount of open canopy in forested areas;
- Creates conditions favorable to wildlife that depend upon ponds, pond edges, dead trees, or other habitats in streams not modified by beaver;
- Increases the mass of insects emerging from the water surface;
- Creates favorable conditions for the growth of bank-stabilizing trees and shrubs, including willow and alder;
- Greatly increases the amount of organic carbon, nitrogen, and other nutrients in the stream channel;
- Ameliorates stream acidity;
- Increases the ecosystem's resistance to perturbation.

As we enter a period of longer droughts, bigger floods, and rising demand for increased water quality and quantity, competition among water users will only increase. Here's one simple answer: get beavers back to work.

To top it off—beavers do their carbon engineering for free!

TO LEARN MORE

"Beaver as a Climate Change Adaptation Tool:
Concepts and Priority Sites in New Mexico."
Available from the Seventh Generation Institute at:
www.seventh-generation.org/resources

"Landscape-scale Carbon Storage Associated
with Beaver Dams" by Ellen Wohl. *Geophysical
Research Letters,* no. 40, 14 (2013), 3631–3636.

Bats Need Water Too

Sharing not fighting

It's getting harder to find a good drink of water in the arid West, especially if you're a bat.

Scientists say the West has been on a drying trend since 2002 and is projected to keep getting hotter and drier well into the future, thanks to climate change. Bat biologist Dan Taylor can confirm the trend from his own experience. Over the course of more than 20 years in the field, Taylor has watched creeks, ponds, and other water sources shrink and decline across the region. Worse, by some estimates 80 to 90 percent of the West's riparian habitats—by far the most important to wildlife—have been destroyed by development or exist in a degraded condition as a result of human mismanagement, including overgrazing by cattle. Although progress has been made recently in restoring riparian habitats, as well as changing grazing practices, Taylor believes that the overall downward trend in water availability will continue, hurting the chances of survival for wildlife and domestic livestock alike.

However, contrary to Mark Twain's famous quip that in the West "whiskey is for drinking and water is for fighting," Taylor and his employer, Bat Conservation International, have found a way for bats and cattle to coexist in hotter and drier times. And not only coexist—but also rely on one another for survival.

It's not just about bats either. Planning for the water needs of a wide variety of wildlife is destined to become a major endeavor in the near future, which means that figuring out win-win solutions to engage agriculture constructively will be crucial. Bats are useful in this regard because their water requirements are very narrow and generally poorly understood by landowners, so if we can get bat water right by working together it can serve as a role model for other cooperative endeavors.

A Townsend's big-eared bat. Bats drink on the fly, requiring a "swoop" zone of a sufficient length and free from obstacles. *Photo by Merlin Tuttle, Bat Conservation International*

Like most mammals, bats need water on a regular basis, especially during hot weather, when they can lose up to 50 percent of their body weight in a single day. Since they don't get enough water from the insects they eat, bats must depend on freely available water for their survival. It must be pooled water too. Bats drink on the fly and thus require a "swoop" zone, just like airplanes do at airports, of a sufficient length and free from obstacles. The depth of the pooled water isn't important, just the access for swooping. Although different bats have different requirements for the size and shape of pooled water, all of them share a critical concern: as the slowest reproducing mammal on the planet for their size, averaging just one pup per year, the loss of reliable water sources can jeopardize an entire colony.

Which is where livestock (and humans) come in.

Hundreds of thousands of water developments for livestock have been put in place across the West since the 1950s, many in the form of stock troughs fed by windmills. But most of these troughs are not bat friendly. Some bat species can maneuver in small spaces, but most need a pool at least 10 feet long and a few require a swoop path 50 to 100 feet long (a river or stock pond) to get a drink. Obstacles such as wire fences and cross braces in the swoop path can prove deadly to a bat in flight. If a bat strikes one and falls into the water, it will drown unless there is an escape ramp provided for it.

For a minimal cost, landowners can make stock troughs bat friendly by maintaining a steady water supply (don't shut off the water when the cows leave); keeping the water's surface as free of obstructions as possible; and providing permanently installed wildlife escape ramps and ladders made from long-lasting material, such as expanded metal.

Other useful sources of water for bats are open-top storage tanks, some of which are 20 feet in diameter. However, if the water level in a tank falls by even a few inches, it becomes a deathtrap for bats—a problem easily solved by installing escape ramps. There are a variety of other strategies that can assist other types of wildlife as well.

"As these livestock water developments increasingly replace or augment diminishing natural sources," said Taylor at a workshop I attended, "they have become crucial for many species, especially when animals are stressed by drought, high temperatures, or rearing young. Without reliable sources of water, wildlife must either leave or die—to the long-term detriment of rangelands and forests."

Bats are essential to both healthy ecosystems and human economies. They pollinate plants and disperse seeds, for example. Some plants, including the wild agave, require bats for pollination and thus for reproduction. No bats, no wild tequila! Bats also eat tons and tons of night-flying insects, including moths, grasshoppers, and crickets. Many of these creatures, including army cutworm moths and leafhoppers, cost American agriculture billions of dollars annually. There are 45 bat species across the US, 25 of which are found in the Southwest. Improving their access to safe watering sources is thus critically important, especially in dry times.

Bat-accessible water also benefits birds that drink in flight, including swifts, swallows, and nighthawks. Pollinators of all sorts like pooled water, too, as do many other wildlife species, from javelina to

A double rainbow over a bat-pond project on public land, near Santa Fe, New Mexico. Of the 45 bat species in the US, 25 live in the Southwest. *Photo by Courtney White*

cougars. Enhancements to dirt tanks and stock ponds are critical as well. The federally listed Chiricahua leopard frog, for example, has come to depend on stock ponds for its survival in certain parts of the Southwest. Of course, enhancing this source of water is beneficial to livestock as well.

In fact, Taylor believes, well-developed stock ponds could be key to climate change adaptation for many species in the arid West.

"Stock ponds capture surface runoff and have been used to water livestock for more than a century," Taylor said. "They've also become an essential source of water for countless species of western wildlife, including big game, birds, bats, other small mammals, and amphibians. But many are dry or degraded today. We can restore them, but do it in such a way that we create a kind of wetland pond, which will be good for all animals."

This restoration involves lining the bottom of the old stock pond with clay soil and compacting it to prevent water leakage. Decreasing slopes and rebuilding spillways can reduce erosion and give the pond a more natural appearance. Installing large woody debris (such as logs) in small coves constructed along the water's edge and then planting those coves with native species creates a diverse habitat for wildlife. Fencing is modified so that cattle have access to the pond at only one small area, which is hardened by gravel or other material to reduce erosion.

"The end result of these improvements is much higher quality water for livestock, more reliable water for livestock and wildlife, and the creation of high quality wetland habitat," said Taylor. "It's a classic win-win, especially as these areas get hotter and drier under climate change."

Helping bats and other wildlife find water means helping ourselves, to the benefit of all.

TO LEARN MORE

For more information about Bat Conservation International, see: www.batcon.org

To read Bat Conservation International's "Water for Wildlife," visit: www.batcon.org/pdfs /water/bciwaterforwildlife.pdf

For the Birds

A novel approach to helping endangered species

Connecting the dots can sometimes lead to unexpected opportunities.

What's the link, for example, between a chicken-like bird with a spiky tail that's struggling for survival; a vast region of high, cold desert shrubland that's been taken over by a tenacious annual grass plant native to Europe; millions of acres of land on fire; and an opportunity to positively influence climate change? The unexpected answer: soil carbon. Thanks to new scientific research, this link has become clear only recently, opening an intriguing and potentially important new approach to endangered species protection, ecological restoration of large landscapes, and rural economic development.

Let's start with the bird, *Centrocercus urophasianus*, or the greater sage-grouse. Once numbering 16 million individuals spread across 14 western states and 3 Canadian provinces, the sage grouse population has dwindled to less than 500,000 on about one-third of its original range, and as a consequence has become a candidate for listing under the Endangered Species Act (ESA). Drought in the region has played a role, but the main reason for the plummeting numbers is habitat fragmentation and loss. This is partly due to housing developments and oil and gas exploration, but mostly it's due to the widespread conversion of landscapes once dominated by sagebrush to ones dominated by an invasive annual grass. This is significant because during winter months, 99 percent of a sage grouse's diet consists of sagebrush leaves. No sage, no grouse.

Not only are sage grouse important biologically in the high, cold desert country of the Interior West, the flamboyant, strutting courtship dance of the male at breeding time makes them a popular attraction for bird watchers.

Once numbering 16 million individuals, the sage grouse population has dwindled to less than 500,000 on about one-third of its original range.
Photo by Tom Reichner/Shutterstock

The attacking annual grass is called cheatgrass, which hitched a ride to the US from Europe in the mid-nineteenth century and eventually came to be cursed as "the invader that won the West." Its success is tied to its competitive advantages over other grass plants: it germinates in the fall, matures in early spring, hogs moisture, and produces a huge amount of seeds (15,000 per square meter, typically) that persist in the soil for years. It loves disturbed soil, which made it the ideal candidate to take over land demolished by overgrazing livestock during the cattle boom years (1880–1920). But what cheatgrass does best is *burn*. By drying out early in the year, it not only becomes a highly combustible fuel; it's also ready to colonize land denuded of sagebrush by fire. It's a devastating cycle: more cheatgrass = more fires = less sagebrush = more cheatgrass. Round and round.

Since 1990, over 20 million acres in the Great Basin have burned (nearly 2 million in 1999 alone). Today, at least 50 million acres of former sagebrush country is now cheatgrass country—in some places to the point of becoming a monoculture. Making matters worse, there is emerging evidence that cheatgrass production is enhanced by elevated levels of atmospheric carbon dioxide.

None of this is good news for the sage grouse, of course, but it's also put the federal agencies overseeing its habitat into a difficult bind. On one side are various wildlife advocacy groups, many of whom are suing to have the bird protected under the ESA, and on the other side are the oil and gas and ranching industries, who are fearful that a listing would impose big restrictions on their activities. For their part, the agencies have explored a variety of sagebrush restoration initiatives over the years, but with limited success so far.

Which brings us to soil carbon.

In a scientific paper, Susan Meyer, a research ecologist with the USDA Shrub Sciences Laboratory in Provo, Utah, wrote that cold desert shrublands are a great place to store carbon in soils for long periods of time. Sage plants have naturally deep rooting systems, creating high organic reserves in the soil. However, these carbon stocks are lost when the land is degraded or converted to cheatgrass, which transforms these systems from net carbon sinks (storing CO_2 in the soil) into net carbon sources (respiring CO_2 back up into the air). Economically, she writes, as carbon sequestration becomes increasingly valued by society, shrubland management for carbon storage could become a source of revenue to accomplish land restoration and sage grouse management goals.

According to Meyer, cold deserts have lots of carbon in their soils, because the roots of sage plants transfer carbon deep underground and also because a lack of water most of the year limits the rate of microbial respiration in the soil (that is, limits the amount of CO_2 rising into the air). These conditions led to the accumulation and persistence of carbon stocks over the years and give them an advantage over other types of dry country. "Because cold deserts store much of their carbon below ground and that carbon is stored in deeper soil layers," Meyer wrote, "these deserts are likely to store more carbon per unit area than warm deserts with monsoonal moisture regimes."

Cheatgrass country is almost the mirror opposite. The plant's shallow root system cycles carbon much more rapidly, the loss of sage plants eliminates deep carbon storage, and frequent fires send lots of earthbound carbon into the atmosphere as carbon dioxide.

Sagebrush country. Nonnative cheatgrass has invaded over 50 million acres of land in the American West, threatening sage grouse habitat. *Photo by marekuliasz/Shutterstock*

Although deserts and semideserts occupy 22 percent of the Earth's land surface, Meyer notes, strategies to mitigate climate change have rarely considered improving carbon sequestration in dry country. This is unfortunate because "improving carbon sequestration in deserts by restoring degraded shrublands to a more functional state would address a broad suite of resource values, including improved air and water quality, wildland fire abatement, enhanced wildlife habitat, biodiversity conservation, and aesthetic and recreational values."

Good news for the sage grouse!

This job won't be easy, especially since the success rate for establishing shrub seedlings is low and will go lower under the hotter and drier conditions predicted by climate scientists. For her part, Meyer is focused on natural pathogens that inhibit cheatgrass growth, including two with vivid names: the Black Fingers of Death and Bleach Blonde Syndrome. Neither is a silver bullet, however.

Which raises a question: What about livestock? A study out of the University of Nevada-Reno demonstrated that when cattle eat cheatgrass in the fall or winter (when it has more protein), there will be less cheatgrass growth in the spring, giving perennials a chance. Sheep will also eat cheatgrass, as will goats, though grass is low on a goat's menu of preferable food.

To this end, the Sage Grouse Initiative, a partnership of ranchers, agencies, universities, and nonprofit groups, is working on solutions for the imperiled bird, and they believe that livestock can help. The key is controlling the timing, intensity, and frequency of grazing to sustain native grasses, wildflowers, and shrubs. And one of the best ways to sustain all of the above is by increasing the carbon stored in the soil.

It's a tall order, of course, but the first step is to dream big.

TO LEARN MORE

For more information on the Sage Grouse Initiative,
see: http://www.sagegrouseinitiative.com

"Restoring and Managing Cold Desert Shrublands for
Climate Change Mitigation" by Susan Meyer. Chapter 2
in "Climate Change in Grasslands, Shrublands, and Deserts
of the Interior American West: A Review and Needs
Assessment" edited by Deborah M. Finch. USDA Forest
Service, Rocky Mountain Research Station, 2012.

Bear with Us

The value of keeping an open mind about predators

Wolves and bears are staging a comeback in crowded, urban Europe, challenging long-established and cherished ideas about conservation.

I'll admit to doing a double take when I first heard the news about a new study published in *Science* that said populations of big carnivores, including bears, wolves, and lynx, are on the rise across Europe—of all places! As a young man growing up in the United States, I had accepted as unshakeable doctrine the Big Wild philosophy that carnivores and other highly mobile wildlife required room to roam—lots and lots of room—specifically national parks, wilderness areas, and other kinds of protected landscapes. Anything less wasn't as good, I was instructed, including agricultural landscapes (especially cattle ranches). Advocates promoted this vision with colorful maps of the US that ranked regions by their wildness—protected public lands merited top billing, with working private lands in the middle, exurban areas next, and cities getting the bottom rank. The message was clear: do everything we can to keep humans and their activities as far away from big, wild animals as possible.

Yet here comes the news that Europe, one of the most trammeled landscapes on the planet, with lots of villages and roads and very few protected areas of any appreciable size, is now home to 12,000 gray wolves (*Canis lupus*). That's twice as many as can be found in the contiguous US, despite the fact that Europe is half the size and more than twice as densely populated.

As in the US, wolves were nearly hunted to extinction on the European continent in the early twentieth century, largely to protect livestock from depredation (though sport hunting by aristocrats played a role as well). Legal protection was key to reversing the wolf's decline,

According to a new study, populations of wild carnivores such as this European lynx are on the rise across urban, densely populated Europe. *Photo by Michal Ninger/Shutterstock*

as it has been in the US, but the next phase of the wolf recovery effort took a different approach in Europe, one that emphasized integrating carnivores among human populations instead of isolating them in parks and conservation areas.

In the *Science* study, a team of more than fifty carnivore biologists from across Europe also researched the status of brown bears (*Ursus arctos*), wolverines (*Gulo gulo*), and the Eurasian lynx (*Lynx lynx*). Their analysis showed that roughly one-third of mainland Europe hosts at least one large carnivore species, and all four species had stable or increasing populations. For example, there are an estimated 17,000 brown bears across 22 nations in Europe, up slightly from previous estimates. Moreover, according to the authors all four large carnivore species persist in human-dominated landscapes and exist largely outside protected areas.

Apparently, wild wolves can be found forty minutes from Rome!

Of course, wild animals need space to be wild in, and in this regard a slow but steady abandonment of marginal farmland across the continent over the past few decades, as food production has consolidated into bigger and bigger operations, has created more elbow room for wildlife. In crowded Europe, this has been an important development. Nevertheless, the study's authors credit other circumstances for the near-record numbers of wolves, bears, wolverine, and lynx.

"The reasons for this overall conservation success," the authors wrote, "include protective legislation, supportive public opinion, and a variety of practices making coexistence between large carnivores and people possible. The European situation reveals that large carnivores and people can share the same landscape."

Perhaps most remarkable has been the profound transformation of public sentiment in support of coexistence with predators. Don't forget, Europe is home to the Big Bad Wolf of Red Riding Hood fame. As the study notes, there has been a deeply rooted hostility in the region to these species throughout its long human history. That attitude, however, has largely been replaced in recent years by one of tolerance, though it's not one necessarily shared by farmers and other rural owners of livestock. Still, in densely populated Europe, it is a crucial change.

"The European model shows that people and predators can coexist in the same landscapes," said Guillaume Chapron, a biologist with the Swedish University of Agricultural Sciences and the study's lead author. "I do not mean that it is a peaceful, loving coexistence; there are always problems. But if there is a political will, it is possible to share the landscape with larger predators."

The study's authors go on to say that if Europe had tried to practice American-style predator conservation, focused on parks and wilderness, there would be hardly any large carnivore populations at all; most European protected areas are too small to host even a few large carnivore reproductive units. This is significant because the two main drivers of the current biodiversity crisis globally—human overpopulation and the overconsumption of resources—show no sign of slowing down, which means that coexistence is more important than ever.

Which is easier said than done, of course.

Recently, a group of French agricultural researchers and agricultural experts wrote an open letter protesting what they see as a swing of the pendulum too far toward wolves, whose high numbers and protected status now threaten the time-honored and eco-friendly practices of the region's pastoralists. Herding, they wrote, is highly respectful of biodiversity and provides a variety of ecosystem services. By its nature, it is also a model of coexistence, and for twenty years herders have employed a variety of strategies to protect their flocks while respecting their predators, including night penning with electric fencing, guard dogs, assistant herders, noisemaking technology, and increased surveillance. Alas, livestock losses have doubled in four years.

"While farmers and herders have modified their practices," wrote the researchers in their letter, "so too have the wolves, and the wolves seem to be winning. Even more worrying, the presence of humans no longer appears to be dissuasive."

The answer, they insisted, isn't a general retreat of agroecological activities in the face of predator pressure, as some pro-wildlife advocates have suggested. Pastoralists also have a right to stand their ground and continue their traditions, the researchers argue. Society needs them, too—small farmers provide local, quality agricultural products while maintaining the vitality of diverse and appealing landscapes. Many members of the public understand this, but what they don't understand is how much the herders' situation is becoming increasingly untenable, put in peril by the rising numbers of ever bolder, and ever popular, predators. It's a serious conundrum.

Expanding populations of predators in Europe create challenges for traditional herding activities, requiring constant vigilance. *Photo by miroslavmisiura/Shutterstock*

"Can we still convince wolves to remain 'wild' by enjoining them to keep their distance from livestock activities?" ask the letter writers, not so rhetorically. In other words, without *shooting or trapping them*?

I don't know the answer, but clearly coexistence means wolves getting along with humans as much as vice versa. Coexistence between humans and all other species on the planet will be vital to our collective future, especially under the rising stress of climate change. It begins with a shift in our attitudes and prejudices away from domination and exclusion toward cooperation and sharing. It's not a utopian vision, as I have tried to illustrate with this story. Coexistence is a complicated business, fraught with unhappy compromises and hard choices, but if we intend to "have our Earth and eat it too," then we had better be prepared to readjust our thinking, as well as our practices.

One acre at a time.

TO LEARN MORE

"Recovery of Large Carnivores in Europe's Modern Human-dominated Landscapes," by Guillaume Chapron et al. *Science,* 346, no. 6216 (2014): 1517–1519.

For more information on carnivore conservation in Europe, see: www.lcie.org

A Burning Question

The thin line between helping and hurting

Wildfire is a keystone ecological process in nature and an ally in our efforts to restore land to health, but it also contributes to global warming, so how do we balance competing goals?

A few years ago, scientists at the National Center for Atmospheric Research (NCAR) and the University of Colorado published a study that said large-scale wildfires in the US generate nearly 300 million metric tons of carbon dioxide a year, which is roughly equivalent to 5 percent of the nation's total annual CO_2 emissions from burning fossil fuels. "Enormous fires pump a large amount of carbon dioxide quickly into the atmosphere," said Christine Wiedinmyer of NCAR, one of the study's authors. "This can complicate efforts to understand our carbon budget and ultimately fight global warming."

Making matters worse, scientists say larger, hotter, and more frequent wildfires are very likely to occur in the future under drier conditions brought on by climate change, especially in the vulnerable evergreen forests of the West and South. The most carbon-laden trees are those with dense wood and large trunk diameters, which are often the victims of the big, hot fires. In 2006, over 95,000 wildfires destroyed about 10 million acres of forest across the United States. Most fires are ignited by lightning, but some are caused by humans—either way, it appears that ever-larger amounts of CO_2 are being released into the atmosphere.

These fires are largely the consequence of a century-long policy of complete fire suppression by the US Forest Service and other landowners, which has resulted in highly overgrown—and flammable—forests. When a fire eventually sweeps through, it often reinvigorates the land, spurring the growth of new vegetation that could ultimately absorb as much CO_2 as was created by the fire. It's a recovery process, however,

Large forest fires can generate CO_2 emissions equivalent to 5 percent of the nation's annual total CO_2 emissions from burning fossil fuels. *Photo by Arnold John Labrentz/Shutterstock*

that can take many decades. And many forests burn so hot now that they never fully recover. The only practical course of action is to prevent small wildfires from becoming big ones.

Fortunately, a solution has already been tried, tested, and perfected: prescribed fire. This is an intentionally set burn that is carefully designed to mimic smaller fires caused by nature or set historically by Native Americans. The goal of a prescribed fire is to remove dead, fallen, and densely crowded wood in a forest in a manner that restores proper ecological functions to the land. Sometimes it is coupled with thinning activities, in which trees are removed mechanically (and often sold commercially) prior to the burn. Either way, the overarching goal is to create conditions for a "cool" fire in the forest, when a fire inevitably comes. By removing underbrush and other flammable material, cool fires are beneficial to forests, unlike the intensely hot and destructive conflagrations that we experience these days, which are often called crownfires because they extend up into the crowns of trees, killing them.

The role of cool fires, say researchers, can be the difference between a forest ecosystem functioning as a carbon sink or a carbon source in the long run.

While prescribed fire has long been viewed by foresters and scientists in a positive light, recent studies suggest it has a productive role to play in climate change as well. Using satellite imagery and computer modeling, scientists at NCAR found that prescribed burns could reduce the carbon emissions of forest fires by an average of 18 to 25 percent and by as much as 60 percent in certain forest systems. That's because prescribed fires release less carbon than wildfires of comparable size, reduce the risk of catastrophic, carbon-spewing wildfires, kill fewer large trees (which store a lot of carbon), and create healthier forests that are better able to sequester carbon than their overstocked cousins. So, the solution is straightforward, right? Light a fire!

Maybe not so fast.

Prescribed fires are difficult to set in remote locations and may not be the right tool in some forested landscapes, due to concern for wildlife, proximity to houses, or other resource constraints. They are also expensive to design and implement, which means they must compete for funds with other budget priorities within agencies. Sometimes there is local community resistance to a deliberately set fire and sometimes a well-intentioned project can become hopelessly bogged down in bureaucratic red tape. Some environmental activists reject prescribed fire outright, usually citing the dangers posed by fire to wildlife or a general "leave nature alone" philosophy. There are even a few researchers who dispute the science, insisting that large, hot fires can be natural too.

Then there's practical concern that a fire might get away from its handlers. I witnessed this firsthand in May 2000, when the National Park Service lit a prescribed fire in Bandelier National Monument on a windy day. The fire escaped its boundaries, blowing up into a major forest fire that forced the evacuation of 20,000 residents from nearby Los Alamos and threatened the nuclear laboratory. In the end, it burned 48,000 acres and became one of the costliest fires in New Mexico's history, in terms of public relations as well as dollars.

"Fires are going to burn in the forests in the western United States," said NCAR's Wiedinmyer. "It's partly up to us to decide how we want that to occur. Carbon is just one piece of the puzzle."

There are lots of risks, in other words, to playing with matches or, in this case, drip torches.

Less controversial is the role fire plays in prairie and other grassland ecosystems. Much less carbon dioxide is released into the atmosphere during a prairie fire than during a forest fire, and generally the amount of carbon emitted is more than offset by carbon stored in grassland soils. (Don't till them, however.) This is even true of annual burning of prairie, whether a consequence of a lightning strike or deliberately set by a land manager. Like grazing, burning can promote grassland vegetative growth if done properly. Old or dead grass and other plants are burned off, making way for new growth. And there's evidence that fire can stimulate soil microbes to make more nitrogen available to plants. These are some of the reasons ranchers and other landowners have embraced prescribed fire as a key element of their land management toolbox.

However, if rain doesn't follow the fire, then the land (and its owner) could be in serious trouble. Cool fires can set the conditions for ecological renewal—but without subsequent moisture, plants won't grow and the land can quickly become susceptible to wind erosion. Drought plus fire can be as lethal to the health of an ecosystem as drought plus

Cool fires like this prairie fire can set the conditions for ecological renewal, but without subsequent moisture, plants won't grow. *Photo by Svitlana Kazachek/Shutterstock*

overgrazing, and just as long lasting. Even a natural fire under these circumstances can be a huge challenge, necessitating that it be put out pronto. Fire always involves a roll of the dice. Its benefits often outweigh its risks, but the risks can be large. One thing is certain: fire is part of nature's plan and it'll happen sooner or later no matter how we feel about it.

To burn or not to burn is a tough question—and emblematic of the difficult choices we face as we move deeper into the twenty-first century.

TO LEARN MORE

"Estimates of CO_2 from Fires in the
United States: Implications for Carbon Management,"
by Christine Wiedinmyer and Jason C. Neff.
Carbon Balance and Management, 2007:
http://www.cbmjournal.com/content/2/1/10

"Sink or Source? Fire and the Forest Carbon Cycle,"
by the Joint Fire Science Program of the US government.
Fire Science Brief, issue 86, January 2010:
http://www.firescience.gov/projects/briefs
/03-1-1-06_fsbrief86.pdf

Connectivity
Interdependence rules

In nature, nothing exists in isolation.

Every living creature, from microscopic bacteria in the soil to blue whales, exists in a web of interdependent relationships, some as prey, some as predators, some as shelter, some as the sheltered. This web of interconnectedness includes humans, though we frequently (and often intentionally) overlook our bonds to the rest of life on the planet. Ironically, while we cherish independence, especially from tyranny, we often ignore the damage we do to the web of interdependence that allows us to pursue life, liberty, and happiness.

One person who understood the interdependence of life was John Muir, a Scottish-born nature enthusiast, amateur geologist, energetic activist, and occasional mystic who dedicated his career to preserving the natural beauty of the high country of California. Passionate and articulate, Muir worked tirelessly to explore and defend the sacred interconnectedness of the natural world. To this end, he founded the Sierra Club in San Francisco in 1891. He also tried to communicate his sense of holism to the public through his many books and articles, one of which included his famous observation that "When we try to pick out anything by itself, we find it hitched to everything else in the Universe."

Unfortunately, the American conservation movement departed from Muir's holistic vision, preferring instead to isolate nature from human use as much as possible—demonstrating in the process how deeply "freedom thinking" runs in our culture. It was assumed by many conservationists that nature could somehow be kept free from our influence and shielded from our poor behavior. Alas, global warming, among many other megaindustrial activities, proved this belief to be an illusion. Muir was right, just not in the way he hoped.

Efforts to restore and protect wild biodiversity require connectivity between wild places and well-managed cultivated landscape, so that animals have room to roam. *Photo by Galyna Andrushko/ Shutterstock*

By the mid-1990s, however, things began to change. Innovative models of sustainable use, collaborative conservation partnerships, and deeper levels of ecological understanding rose and began to spread, watershed by watershed. Negative attitudes toward livestock, for instance, shifted with the emergence of a style of ranch management that mimicked natural patterns of wild herbivores. Simultaneously, the rise of regenerative agricultural practices, environmental restoration projects, and other new tools contributed to an emerging holistic vision of healthy grass, soil, water, animals, people, and local economies, much of it centered on the production of food.

A good example of this change is the Wild Farm Alliance, a non-profit based in Watsonville, California. Founded in 2000 by a group of wilderness proponents and ecological-farming advocates, its goal is to explore the common ground between the production of healthy food and the protection and restoration of wild biodiversity. I first came across this concept in 2003 in a book by Daniel Imhoff titled *Farming with the Wild* that described real-world examples of ecologically managed farms and ranches that integrated a wide range of native plants and animals into their work. I thought it sounded a lot like the Quivira Coalition! In fact, Jim Winder, a holistically minded rancher and Quivira cofounder, was one of the people described in the book.

The organization's work is important because 66 percent of all land in the lower 48 states is in agriculture, to one degree or another, while only 5 percent is federally protected as parks or wilderness areas. This means wildlife populations are vulnerable to habitat destruction and fragmentation, as well as water pollution, pesticides, and other effects of industrial food production. Furthermore, the majority of endangered species in the US are found on private land, much of it owned or managed by farmers and ranchers. As a result, conservation efforts on behalf of wildlife must engage the agricultural community in order to succeed. At the same time, in an increasingly urbanized world farmers and ranchers need city-based conservationists as allies.

The Wild Farm Alliance is trying to bridge these divided worlds by reconnecting food systems with ecosystems.

The organization accomplishes its goals in two ways. The first is the promotion of farming practices that help restore and maintain wild habitat and native species, including: farming without toxic chemicals or genetically modified organisms; growing locally adapted crops and animals; managing livestock with intensive or planned grazing strategies; raising grassfed animals; restoring native perennial forage; using cattle, goats, and sheep to eat invasive weeds; periodic prescribed burning; protecting and restoring riparian areas; and planting hedgerows, shrubs, and trees for native species.

On its website, the Alliance notes that suitable habitat for livestock grazing encompasses 40 percent of the land base in America (excluding Alaska and Hawaii). If farms and ranches adopted regenerative practices, the potential positive impact on wildlife and other forms of biodiversity could be huge. "When optimally managed," says an Alliance paper, "farms and ranches support healthy grasslands and wildlife forage plants, efficient watersheds, significant areas of habitat, wildlife connectivity, and buffering for wildlands."

The Alliance's second goal is the protection and restoration of wildlife-movement corridors between islands of wild country, complemented by ecologically managed farms, ranches, and forests. The aim is to reduce any further fragmentation of natural habitat while reconnecting pockets of protected land. Typically, corridors are narrow strips of unobstructed land, which allow wildlife room to find water, forage, shelter, and mates. A creek is a good example of a corridor, stretching as it does through a watershed. When free from dams, fences, or other human obstructions, a creek can serve as a natural conduit for migrating birds, fish, and mammals.

Corridors can be big like a ranch, linking large landscapes, or they can be small like a passageway over or under a freeway. They can come in any shape and serve any type of animal. But they all share a common purpose: to ensure connections between isolated patches of habitat so that one or more species can move freely back and forth.

Particularly important are links between public and private land, especially if a farm or ranch connects two biologically significant blocks of public land. Too often historically, however, the management philosophies of these separate types of land have clashed with one another. Fortunately, in recent years private and public landowners have come to understand that they need each other if things are going to work on larger scales, and a good-faith effort is underway to find common ground.

Wildlife corridors, such as this newly constructed highway overpass, help to ensure that animals can travel between isolated patches of habitat. This is one small example of a way to restore connectivity to promote ecosystem health. *Photo by Pics-xl/Shutterstock*

There's more to connectivity than map making and best management practices. As the Muir quote suggests, it has a metaphorical role too. Do we live in separate boats, rowing in different directions, or do we exist in one boat? Are we hitched together, bound by a common cause, or are we divided? In the mid-twentieth century, the answer appeared clear: we could make it in separate rowboats. Today, however, it's obvious that we're all in the same boat, drifting toward the same large rapids. In the interest of all life on the planet, shouldn't we row together? We can, I believe, especially if we understand that aboveground health is inextricably linked to soil health. Health flows upward from the soil. It doesn't matter whether we're talking about wildlife, forests, butterflies, livestock, food, or ourselves, it's all a web of interdependence.

The sooner we grasp this fundamental fact of nature the better!

TO LEARN MORE

For more information about the Wild Farm Alliance, see: www.wildfarmalliance.org

For an example of large-scale corridor connectivity, see the Yellowstone to Yukon Conservation Initiative website: www.y2y.net

Farming with the Wild: Enhancing Biodiversity on Farms and Ranches by Dan Imhoff. Sierra Club Books, San Francisco, CA, 2003.

Positive Change

Finding wisdom by observing animal behavior

I t's a fact of life that we live in a world of constant change, but how many people actually accept the inevitability of change? Not many, I bet.

"Change is never painful," the Buddha said a long time ago, "only the resistance to change is painful." Most people, myself included, like things the way they are generally and don't often stray out of our comfort zones. While some are more willing than others to explore new ideas, many cling stubbornly to old beliefs and habits, which is how human nature operates, I suppose. But tell it to the world, which is constantly moving on and now, in the twenty-first century, threatens to move *way on*. There is little doubt that this is going to be a century of unprecedented change, raising important questions about how much painful resistance we are going to try to throw in its path.

In this regard, we could learn a lesson from nature, particularly the animal world, says Fred Provenza, an emeritus professor of behavioral ecology at Utah State University.

Growing up in southern Colorado and working on area ranches, Provenza became fascinated by the behavior of sheep, cattle, deer, and elk on the open range. He was especially curious about why animals chose to eat what they did, observing that people who make a living in ranching often ignored "how animals make their living," as I heard him say in a lecture. Provenza pursued his curiosity through a master's degree in range and wildlife science, a PhD in animal behavior, and a long career of research. What Provenza learned is that we are all creatures of habit for a reason.

Take nutrition. Herbivores eat a diverse array of plant species, as many as one hundred different ones, but studies showed that the bulk of any particular herbivore meal normally contains less than ten plant

Humans can profit materially and spiritually from observing the nature of animals, including domesticated herbivores. *Photo courtesy of Michel Meuret*

species and typically as few as three to five. Although scientists knew that food selection by herbivores is guided by nutrients and toxins in foods, they were surprised to learn how much influence mothers have over foraging behavior, Provenza said. Studies demonstrated that herbivores are nutritionally wise, a conclusion that contradicted the long-standing belief that herbivores are generally "unwise" because they don't always choose the most nutritious foods to eat. This belief created a paradox: we are often baffled when livestock don't do well despite an abundance of suitable habitats and nutritious forage.

Sound familiar to human nutritional problems?

One key to resolving this paradox, Provenza learned, is a clearer understanding of the role experiences early in life play on shaping diet and habitat selection behavior in creating locally adapted animals who don't perform well when moved to unfamiliar environments. Another key is understanding the role toxins play in animal diet and the regulation they require on food intake, a role that influences behavior. By setting limits on the intake of any one food, toxins force animals to eat a variety of foods to meet their nutrient needs. Moreover, every individual is different in its nutrient needs and its ability to cope with toxins. Thus, grazing practices that allow the individuality of animals to be expressed are likely to improve performance of the herd.

Another insight is how animals learn. Provenza and his colleagues discovered that when young herbivores are encouraged to eat only the most preferred plants, they are not likely to learn to mix foods high in nutrients with foods that contain toxins. Experienced animals learn to eat a variety of foods, even when more nutritious foods are available.

Other insights that Provenza and fellow researchers gained include:

- Since life exists at the boundary between order and chaos, animals, humans included, learn habits to create order and predictability;
- The origins of animal food habits and habitat preferences involve interactions between the social organization (culture) of the herd and the individual;

- Although both people and herbivores strive for order, they also seek variety;
- Ongoing changes in social and physical environments require old dogs to learn new tricks all the time.

"Thus, while the behavior of herbivores may appear to be little more than the idle wanderings of animals in search of food and a place to rest," Provenza said, "foraging is a process that provides insights into an age-old dilemma faced by herbivores and humans alike: how do creatures of habit survive in a world whose only habit is change?"

In other words, if we are, as Aristotle once remarked, "what we repeatedly do" then how do we break destructive habits and manage for long-term sustainability?

The first step, according to Provenza, is to try to understand what part of behavior is cultural (habit) and what is not. Take, for instance, livestock grazing in riparian zones. Cattle are not genetically preprogrammed to wallow in creeks. Instead, it is a learned behavior, a habit that can be changed. He often cites the example of rancher Ray Bannister, who manages cattle on his property in eastern Montana according to an extreme version of planned grazing principles, which requires intensive soil- and plant-stressing periods of heavy grazing followed by two years of complete rest. This system forces Bannister's cattle to eat all the forage in a pasture, not just the "ice cream" plants, thus eliminating the competitive advantage of the unpalatable plants.

As a result, said Provenza, "It is hard to find any part of the ranch that lacks abundant plant cover, even during years of drought."

Bannister's challenge, however, was convincing his cattle to change their eating habits. It took three years for his animals to adjust, during which their weight and performance dropped dramatically—but eventually recovered. Now the mother cows teach their calves the system and all is well on the Bannister ranch. In fact, Provenza believes that management-intensive systems can balance animal, plant, social, and economic concerns.

In contrast, humans rely too much on technology and not enough on the culture inherent to social animals, he believes, in particular the collective knowledge and habits acquired and passed from generation to generation about how to survive in a particular environment or a time of change.

If we instead allow cultures to develop along natural lines, we may lessen our dependency on technological fixes and come to rely more on behavioral solutions that cost very little to implement and are easily transferred from one situation to the next. Unfortunately, said

Rainbow at sunset over regenerative farm in New South Wales, Australia.
Photo by Courtney White

Provenza, scientists and managers often ignore the power of behavior to transform systems, despite compelling evidence. Once mastered, he argues, behavioral principles and practices provide an array of solutions to the problems people face today.

Provenza states the general problem this way: "How does one manage ongoing interrelationships among facets of complex, wholly interconnected, poorly understood, ever-changing ecological, cultural, and economic systems in light of a future not known and not necessarily predictable, in ways that will not diminish options for future generations?"

The best place to look for an answer is in nature— and not just with what's "out there" but with what's inside us as well. We are what we do—and we do what we are. Creating viable options for the future requires that we embrace change on both fronts.

TO LEARN MORE

For more information about
Utah State University's BEHAVE program, see:
www.extension.usu.edu/behave

A fascinating lecture by Fred Provenza titled
"The Web of Life" can be viewed at:
https://www.youtube.com/watch?v=ZjUgX91VZpk

Completing the Circle
Restoring wildness, ranchland, and ourselves

The poet T. S. Eliot once observed that at the end of our exploring we will "arrive where we started and know the place for the first time"—which is an apt description of this moment in history.

More than a century and a half ago, most conservationists were hunters, and the conservation movement began its good work out of a concern for wildlife and the threats birds and other animals faced from human activity. The response in the beginning largely centered on isolated refuges and other types of protected landscapes. Over the ensuing decades, as various conservation strategies and philosophies rose and fell, the movement never lost its focus on wild animals, especially ones in danger of extinction. What did change over the years, however, was the way wildlife habitat has been managed, especially on private land.

A leave-nature-alone, hands-off approach is being replaced by nature-knows-best, hands-on practices, to great effect. The conservation desire is the same, but what we can accomplish on the land is very different.

The Gill family, owners of the Circle Ranch, located near Van Horn, Texas, are a good example.

"We cannot restore biodiversity by destroying biodiversity" is how Christopher Gill sums up an important lesson learned from his family's efforts to restore wildlife habitat and

The Circle Ranch. The Gill family uses planned grazing by cattle, Keyline contour plowing, and gully repair and water harvesting to restore the land to health. *Photo courtesy of Circle Ranch*

manage the delicate ecosystem on the Circle, a 32,000-acre slice of high desert in the Sierra Diablo range of west Texas.

In an all-too-familiar story, the Circle endured more than a century of hard use, especially year-round grazing by livestock, which had depleted the ranch's vegetative vigor, encouraged brush encroachment, provoked widespread erosion, and generally made a mess of wildlife habitat, decreasing biodiversity. When the Gills purchased the ranch in the 1990s, they decided to try to reverse these trends. Like many landowners in Texas, they focused their energy and resources on improving the prospects for game animals.

When they started, the Gill family set three broad goals:

- Increase the quantity, quality, and diversity of the ranch's plant community;
- Increase the numbers of free-ranging wild animals, including mule deer, elk, desert bighorn sheep, pronghorn, quail, dove, and turkey;
- Increase the ranch's profitability.

They knew that each goal was interdependent with the others, especially in a state like Texas, which is 98 percent privately owned and relies heavily on market-based incentives such as hunting to encourage conservation work. Unlike other ranches, however, the methods the Gills are using to accomplish these goals are not at all typical.

They employ three primary strategies—planned grazing, Keyline contour plowing, and gully repair and water harvesting—all of which increase life generally on the ranch, in contrast to business-as-usual practices that reduce life, such as overgrazing, trapping, spraying, and poisoning. Life begets life, Gill will tell you, and if you want more wildlife, it's best to start at the level of soil, grass, and water.

Enter herbivores.

"Our primary habitat management tool is cattle," Gill said. "Animal impact and grazing timing are key to our efforts to improve habitat for all species of birds and animals."

As with most "wildlifers," as Gill put it, the family had originally decided not to run cattle on the ranch, believing that livestock and wildlife competed for resources. However, after implementing a planned cattle-grazing program, not only did Gill see a positive ecological result in the form of increased plant vigor and biodiversity, he also saw the advantage of thinking holistically—to see systems as integrated wholes, not as a bunch of specialized or disconnected parts. Wildlife and cattle are linked together by the water, mineral, carbon, and sunlight cycles that make land healthy for both. Habitat can be improved when cattle are used as proxies for wild grazers, mimicking their concentrated numbers, constant movement, and long periods of absence. For Gill, planned grazing is about getting animals to the right place at the right time for the right reason and with the right behavior.

"My conclusion is that cattle offer us a tool that cannot be replicated by machines, chemicals, or fire in terms of the treatment's physiological outcomes," Gill said.

It was profitable too. According to Gill, the Circle Ranch netted $50,000 from its cattle operation in 2013. Any combination of machines, poisons, chemicals, and fire to treat the 20,000 acres that were grazed would have cost at least $30 an acre. So, instead of spending $600,000, the Gill family made $50,000 and also netted an ecological outcome that could not have been created by the more technological practices. Add hunting fees into the economic mix and you have a recipe for a profitable enterprise.

Another practice getting great results at the Circle is Keyline contour plowing, particularly in areas where the tool of animal impact has not worked as quickly as hoped. Using a Yeomans Plow to slice deep, narrow furrows placed on contour by laser transit allows water to gently infiltrate the soil with minimal damage to plants and minimal subsoil disturbance. This encourages plants to reestablish themselves and grow abundantly, and the concurrent root expansion converts subsoil to topsoil. These new and reinvigorated plants can then be maintained by planned livestock grazing. It's a specific tool for a specific place, Gill said, and combined with other management tools it's successfully helping the family achieve their holistic goals.

However, there are jobs that cows and plows can't do, such as repair eroded washes, gullies, and roadbeds, many of which have become open wounds on the ranch, Gill said, "eating whole valleys."

Next up were gully repair and water harvesting.

Implementing methods pioneered by restoration specialist Bill Zeedyk, who encourages landowners to "think like a creek," the Gills have built a variety of structures across the Circle that redirect water flow, slow down flood events, and "re-wet" sweet spots that had dried up due to falling water tables. They have also redesigned ranch roads in order to harvest water falling upslope, redistributing it downslope to grow more grass rather than have it trapped in roadside ditches and shunted away.

Of course, it has to rain. As Gill noted, it would be unrealistic to expect planned grazing, Keyline plowing, gully repair, or any other range practice to work in the absence of rain. However, since it will rain sooner or later, the challenge is to make sure that the water cycle

Keyline plowing involves creating deep, narrow furrows placed on a contour that allows water to gently infiltrate the soil and stimulate the growth of grass for wildlife. *Photo courtesy of Owen Hablutzel*

is as effective as possible. For example, if ten inches of rain falls on a plot of land and eight inches runs off due to degraded or bare soils, the effective rainfall is two inches. On the other hand, if only six inches of rain falls in a drought year and 66 percent of it is soaked up by healthy land, the effective rainfall is doubled to four inches. That's a huge difference in dry country like west Texas.

Gill calls a less effective water cycle brought on by poor land management a "human-caused drought." When combined with a natural drought, the result can be devastating to all life. In contrast, he also calls the practices employed on the Circle "drought busters." More effective water means more grass, which means more wildlife, which means more biodiversity in general. However, don't expect miracles, Gill warned, and don't be in a hurry. If a landowner is willing to be patient, he says, persistence will be rewarded, as it has been for the Gill family. The Circle, once broken, is nearly whole again.

Which brings us back to the place where we started.

TO LEARN MORE

For more information about the
Circle Ranch, see: www.circleranchtx.com

For other writing about restoring land
health, carbon ranching, and related topics
by Courtney White, see: www.awestthatworks.com

Acknowledgments

This book would not have been possible without the Quivira Coalition.

While working on the first chapter of my previous book, *Grass, Soil, Hope*, I wrote a one-thousand-word profile of a successful effort to double the amount of carbon in the soil on the McEvoy Olive Ranch, in northern California. I had intended to include the text in the book, but it didn't fit the flow of the narrative ultimately, so I pulled the profile out, created a Word file, and orphaned the story to a new folder. Later, I came across photographs I had taken during a visit to the McEvoy Ranch with my family. Looking them over, two images jumped out as perfect illustrations for the text I had written. A lightbulb flickered on in my mind: the profile could stand alone. I sent the photos and the text to Avery Anderson and Tamara Gadzia at Quivira for their opinion. Could the organization use a series of short case studies for some publishing purpose? "Yes!" they said. So I kept going, culling profiles from previous work and writing new ones, calling them *2% Solutions*. Eventually, Quivira published 14 case studies as a special edition of its journal *Resilience* in September 2013.

Liking what they read, friends and colleagues in the Quivira community encouraged me to keep writing. Avery and Tamara in particular were strongly supportive and together we came up with the idea of publishing a collection of these *2%* case studies as a book. The idea was endorsed by the staff and board of Quivira, which set wheels in motion. I made a plan, crafted a budget, and set to researching, writing, and fund-raising. Tamara did a brilliant job (as usual) of laying out the case studies as I wrote them and gave me important feedback on content, topics, photo selection, and design. I am especially indebted to Kit Brewer, administrative assistant at Quivira, who diligently read every case study that I produced, as well as the final draft version of this book, providing me with detailed and valuable editorial comments.

In the meantime, we had developed a strong relationship with Ben Watson and Chelsea Green Publishing. When we showed them our draft version of the book in July 2014, they offered to pick up the title and publish it themselves—we were honored! They asked that I write additional profiles so the book could reach a more salable size, which was an assignment that I gladly accepted. The book that you are reading now is the result and it would not have been possible without Chelsea Green.

The book would also not have been possible without financial support of Quivira donors, including the Stokes Foundation, New Cycle Foundation, McBean Foundation, Compton Foundation, Panta Rhea Foundation, Lia Fund, Harry and Barbara Oliver Foundation, Mead Foundation, Healy Foundation, and the Better Tomorrow Fund of the Rockefeller Family Foundation, as well as Chelsea Green Publishing and the board, staff, and members of the Quivira Coalition.

Different versions of a number of the case studies here were previously published in *Revolution on the Range* (Island Press), *Grass, Soil, Hope* (Chelsea Green), *The Age of Consequences* (Counterpoint Press), *Farming* magazine, *Acres*, and various Quivira Coalition publications. I am grateful to the publishers and editors for their help and encouragement.

I also want to thank the broader Quivira community. For nearly 20 years, I've had the deep honor of getting to know an exceptional group of practitioners, thinkers, writers, innovators, teachers, students, and others. Their unfailing goodwill, energy, and innovation have been a vital source of motivation and stimulation for me over the years. Their good work is the real reason this book exists. My main job here is storytelling, but without something exciting to write about there would be no story to tell! I have learned and benefitted in so many different ways from the Quivira community—this book is a small token of my gratitude.

I also thank my family—Gen, Sterling, and Olivia—once again for their love and support. Writing doesn't happen in isolation, despite how it feels after a long day in front of the computer. It takes place within a web of loving, reinforcing relationships. I could not have done it without them!

Lastly, I want to thank every person in this book. Solutions do not happen spontaneously or easily; they are the product of a long process of creation, design, testing, and implementation. The passion, dedication, and hard work of every person mentioned here is astonishing to me—and deeply hopeful. If I have done even a halfway decent job of representing their work, I'll consider my job complete.

—COURTNEY

Index

About the Author

Photo by Elaine Patarini

A former archaeologist and Sierra Club activist, Courtney White dropped out of the "conflict industry" in 1997 to cofound the Quivira Coalition, a nonprofit dedicated to building bridges between ranchers, conservationists, public-land managers, scientists, and others around practices that improve land health (see www.quiviracoalition.org). Today, his writing and conservation work focus on building economic and ecological resilience for working landscapes, with a special emphasis on carbon ranching and the new agrarian movement.

Courtney is the author of *Grass, Soil, Hope: A Journey through Carbon Country* (Chelsea Green, 2014), *The Age of Consequences: A Chronicle of Concern and Hope* (Counterpoint Press, 2015), *Revolution on the Range: The Rise of a New Ranch in the American West* (Island Press, 2008), and an online book of black-and-white photographs titled *The Indelible West* (2012), which includes a foreword by Wallace Stegner. He coedited, with Rick Knight, *Conservation for a New Generation: Redefining Natural Resources Management* (Island Press, 2008).

Courtney's writing has appeared in *Farming, Acres* magazine, *Rangelands, Natural Resources Journal*, and *Solutions*. His essay "The Working Wilderness: A Call for a Land Health Movement" was published by Wendell Berry in 2006 in his collection of essays titled *The Way of Ignorance*. In 2010, Courtney was given the Michael Currier Award for Environmental Service by the New Mexico Community Foundation. In 2012, he was a writer-in-residence at the Ucross Foundation in northeastern Wyoming, and he was the first Aldo Leopold Writer-in-Residence at Mi Casita, the Aldo Leopold cabin in Tres Piedras, New Mexico, courtesy of the Aldo Leopold Foundation and the US Forest Service. He lives in Santa Fe, New Mexico, with his family and a backyard full of chickens. Visit Courtney's website, www.awestthatworks.com, to learn more about his work.

green press
INITIATIVE

Chelsea Green Publishing is committed to preserving ancient forests and natural resources. We elected to print this title on paper containing at least 10% post-consumer recycled paper, processed chlorine-free. As a result, for this printing, we have saved:

13 Trees (40' tall and 6-8" diameter)
6,282 Gallons of Wastewater
6 million BTUs Total Energy
420 Pounds of Solid Waste
1,158 Pounds of Greenhouse Gases

Chelsea Green Publishing made this paper choice because we are a member of the Green Press Initiative, a nonprofit program dedicated to supporting authors, publishers, and suppliers in their efforts to reduce their use of fiber obtained from endangered forests. For more information, visit www.greenpressinitiative.org.

Environmental impact estimates were made using the Environmental Defense Paper Calculator. For more information visit: www.papercalculator.org.

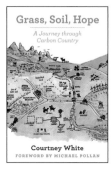

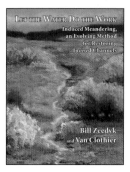
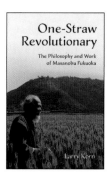

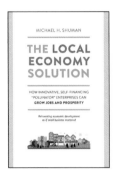

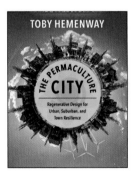